Good Manufacturing Practices for Pharmaceuticals

GMP in Practice

B.N. Cooper

ISBN-13: 978-1974006328

ISBN-10:1974006328

Chapter 1

Overview of Good Manufacturing Practices

Chapter 2

Quality Management

Chapter 3

Personnel

Chapter 4

Buildings and Facilities

Chapter 5

Process Equipment

Chapter 6

Documentation and Records

Chapter 7

Materials Management

Chapter 8

Rejection and re-use of materials

Chapter 9

Validation

Chapter 10

Change Control

Chapter 11

Complaints and recalls

Glossary

"Coming together is a beginning. Keeping together is progress. Working together is success."

Henry Ford

CHAPTER 1

Overview of Good Manufacturing Practices (GMP)

Introduction

Good Manufacturing Practices are a set of practices that are required in order to comply with industry standards and regulations. GMP helps to minimise the risks involved during manufacturing and helps to ensure products meet quality and regulatory standards. A GMP quality system ensures that products are consistently produced and controlled according to predefined quality standards. It is designed to minimise the risks involved in any pharmaceutical production that cannot be eliminated through testing the final product.

CGxP /cGMP

Often, a broader term is used in industry -GxP-where the "x" is used as an umbrella letter representing different subjects or disciplines in industry. Some prime examples include GLP (Good Laboratory Practice), GDP (Good Documentation Practice), GEP (Good Engineering Practice) and GMP (Good Manufacturing Practices). Furthermore, the use of a lower case "c" as a prefix indicates "current" or "up-to-date". So cGMP stands for "Current Good Manufacturing Practices.

This means that some conventions or practices are subject to change within the industry. Therefore, it is important to be up-to-date in the application of cGxP or cGMP

There are multiple regulators and organisations that provide definitions of "Good Manufacturing Practices". They include Organisations such as the World Health Organisation (WHO) and the International Society of Pharmaceutical Engineering (ISPE), PIC/s, EU Eurdralex Volume 4, Good Manufacturing Pracatices Other definitions are offered by bodies such as the American competent authority for Food and Drug Administration. It is good to have an awareness of how organisations, bodies and competent authorities define GMP, and one should always review the "local" regulatory landscape. Below some definitions are provided to provide a feel for GMP and highlight the common thread between definitions.

W.H.O. World Health Organisation-"Good Manufacturing Practices (GMP, also referred to as 'cGMP' or 'current Good Manufacturing Practice') is the aspect of quality assurance that ensures that medicinal products are consistently produced and controlled to the quality standards appropriate to their intended use and as required by the product specification."

Food and Drug Administration: cGMP refers to the Current Good Manufacturing Practice regulations enforced by the US Food and Drug Administration (FDA). cGMPs ensure systems are properly designed and monitored, safeguarding the control of manufacturing processes and facilities. Adherence to the cGMP regulations ensures the identity, strength, quality, and purity of drug products by requiring that manufacturers of medications adequately control manufacturing operations. This includes establishing strong Quality Management Systems, obtaining appropriate quality raw materials, establishing robust operating procedures, detecting and investigating product quality deviations and maintaining reliable testing laboratories. This formal system of controls at a pharmaceutical company, if adequately put into practice, helps to prevent instances of contamination, mix-ups, deviations, failures and errors. This assures that drug products meet their quality standards.

MHRA (Medicines and Healthcare Products Regulatory Agency) defines GMP as follows:

"Good Manufacturing Practice (GMP) is that part of quality assurance which ensures that medicinal products are consistently produced and controlled to the quality standards appropriate to their intended use and as required by the marketing authorisation (MA) or product specification. GMP is concerned with both production and quality control. Many of the drivers of GMP in effect are also benefits to the manufacturer. Good manufacturing practices are an expected practice in regulated industries and a manufacturer must meet all relevant GMP regulations if they wish to manufacture within a country or sell to a particular market. It is important to maintain accurate, complete, up-to-date and consistent information to ensure patient safety and reduce any potential risks."

A basic tenet of GMP is that (1) quality cannot be tested into a batch of product and (2) quality must be built into each batch of product during all stages of the manufacturing process.

Good Manufacturing Practice (GMP) describes a set of principles and procedures that when followed helps ensure that therapeutic goods are of high quality.

There are different codes of GMP, depending on the type of therapeutic good:

➢ Good Manufacturing Practice for Medicines
➢ Good Manufacturing Practice for Human Blood and Tissues
➢ A different system, known as conformity assessment, is used to ensure that medical devices are of high quality.

The following section shows the structure and key headings as they appear in the respective sources. GMP requirements are based on both regulatory authorities and other international organisations such as PICS/s, WHO, FDA etc.

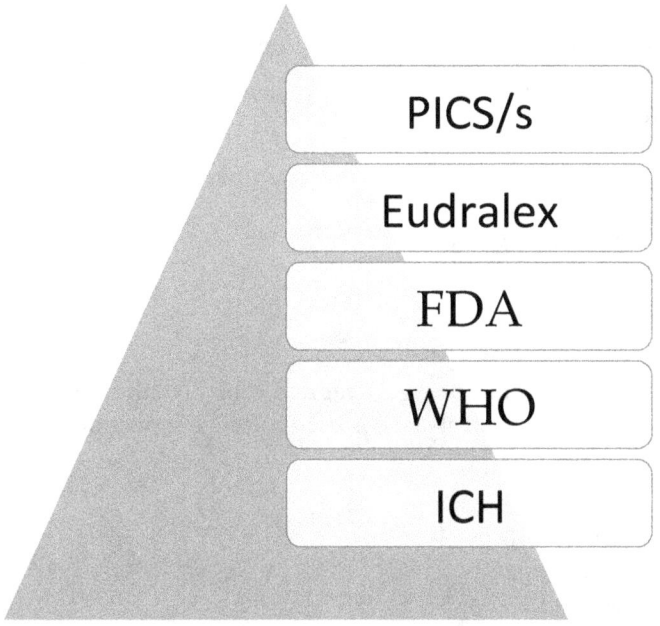

Figure 1: Organisations and bodies that publish GMP requirements

PICS/s manufacturing principles for medicinal products:

Pharmaceutical Inspection Convention and Pharmaceutical Inspection Co-operation Scheme (PIC/S): The Pharmaceutical Inspection Convention and Pharmaceutical Inspection Co-operation Scheme (jointly known as PIC/S) develop international standards between countries and pharmaceutical inspection authorities, to provide a harmonised and constructive co-operation in the field of Good Manufacturing Practices.The PIC/S provides an active and constructive cooperation in the field of GMP and related areas. The purpose of PIC/S is to facilitate:

➢ networking between participating authorities
➢ maintenance of mutual confidence
➢ exchange of information and experience
➢ Mutual training of GMP inspectors.

The Guide consists of an Introduction section along with two parts and a number of annexes.

• **Guide to Good Manufacturing Practice for Medicinal Products** – Introduction
  o Introduction
  o Adoption and entry into force
  o Revision history

- **Guide to Good Manufacturing Practice for Medicinal Products** - Part I
  Part I covers GMP principles for the manufacture of medicinal products

  1. Quality management
  2. Personnel
  3. Premises and equipment
  4. Documentation
  5. Production
  6. Quality control
  7. Contract manufacture and analysis
  8. Complaints and product recall
  9. Self-inspection

- **Guide to Good Manufacturing Practice for Medicinal Products** - Part II
  Part II covers GMP for active substances used as starting materials

  1. Introduction
  2. Quality management
  3. Personnel
  4. Buildings and facilities
  5. Process equipment
  6. Documentation and records
  7. Materials management
  8. Production and in-process controls
  9. Packaging and identification labelling of APIs and intermediates
  10. Storage and distribution
  11. Laboratory controls
  12. Validation
  13. Change control
  14. Rejection and re-use of materials
  15. Complaints and recalls
  16. Contract manufacturers (including laboratories)
  17. Agents, brokers, traders, distributors, repackers and relabellers
  18. Specific guidance for APIs manufactured by cell culture / fermentation
  19. APIs for use in clinical trials
  20. Glossary

The annexes provide detail on specific areas of activity and are listed below:

- **Technical interpretation of PIC/S GMP guide Annex 1 -** Manufacture of sterile medicinal products

  PIC/S has published a recommendation for the technical interpretation of Annex 1 on the Manufacture of Sterile Medicinal Products.

This recommendation summarises the interpretations an inspector adopts during an inspection of the manufacture of sterile medicinal products. It reflects the most important changes introduced in the revised Annex 1, but is not intended to address all changes in the revision.

- o Document history
- o Purpose and scope
- o Basics
- o Definitions and abbreviations
- o New texts and their interpretation
- o Revision history

- **Guide to Good Manufacturing Practice for Medicinal Products – Annexes**

  - o Annex 1 - Manufacture of sterile medicinal products
  - o Annex 2 - Manufacture of biological medicinal products for human use
  - o Annex 3 - Manufacture of radiopharmaceuticals
  - o Annex 4 - Manufacture of veterinary medicinal products other than immunologicals
  - o Annex 5 - Manufacture of immunological veterinary medical products
  - o Annex 6 - Manufacture of medicinal gases
  - o Annex 7 - Manufacture of herbal medicinal products
  - o Annex 8 - Sampling of starting and packaging materials
  - o Annex 9 - Manufacture of liquids, creams and ointments
  - o Annex 10 - Manufacture of pressurised metered dose aerosol preparations for inhalation
  - o Annex 11 - Computerised systems
  - o Annex 12 - Use of ionising radiation in the manufacture of medicinal products
  - o Annex 13 - Manufacture of investigational medicinal products
  - o Annex 14 - Manufacture of products derived from human blood or human plasma
  - o Annex 15 - Qualification and validation
  - o Annex 16 - Qualified person and batch release
  - o Annex 17 - Parametric release
  - o Annex 18 - GMP guide for active pharmaceutical ingredients (This Annex no longer exists)
  - o Annex 19 - Reference and retention samples
  - o Annex 20 - Quality risk management
  - o Glossary

EudraLex - Volume 4 - Good Manufacturing Practice (GMP) guidelines

Volume 4 of "The rules governing medicinal products in the European Union" contains guidance for the interpretation of the principles and guidelines of good manufacturing practices for medicinal products for human and veterinary use laid down in Commission Directives 91/356/EEC, as amended by Directive 2003/94/EC, and 91/412/EEC respectively.

Eudralex V4 is made up of the following parts:

- ➢ Introduction
- ➢ Part I - Basic Requirements for Medicinal Products
- ➢ Part II - Basic Requirements for Active Substances used as Starting Materials
- ➢ Part III - GMP related documents

Introduction

The Commission Directive 2003/94/EC, of 8 October 2003, set out the principles and guidelines of good manufacturing practice in respect of medicinal products for human use and investigational medicinal products for human use.

Part I - Basic Requirements for Medicinal Products

Chapter 1 - Pharmaceutical Quality System
Chapter 2 - Personnel
Chapter 3 - Premise And Equipment
Chapter 4 - Documentation
Chapter 5 - Production
Chapter 6 - Quality Control
Chapter 7 - Outsourced Activities
Chapter 8 - Complaints And Product Recall
Chapter 9 - Self Inspection

Part II - Basic Requirements for Active Substances used as Starting Materials

Basic requirements for active substances used as starting materials

Part III - GMP related documents

Site Master File
Q9 Quality Risk Management
Q10 Note for Guidance on Pharmaceutical Quality System
MRA Batch Certificate

Annexes

Annex 1-Manufacture of Sterile Medicinal Products

Annex 2- Manufacture of Biological active substances and Medicinal Products for Human

Annex 3- Manufacture of Radiopharmaceuticals

Annex 4- Manufacture of Veterinary Medicinal Products other than Immunological Veterinary Medicinal Products

Annex 5- Manufacture of Immunological Veterinary Medicinal Products

Anne 6- Manufacture of Medicinal Gases

Annex 7- Manufacture of Herbal Medicinal Products

Annex 8-Sampling of Starting and Packaging Materials

Annex 9-Manufacture of Liquids, Creams and Ointments

Annex 10-Manufacture of Pressurised Metered Dose Aerosol Preparations for Inhalation

Annex 11-Computerised Systems

Annex 12-Use of Ionising Radiation in the Manufacture of Medicinal Products

Annex 13-Manufacture of Investigational Medicinal Products

Annex 14-Manufacture of Products derived from Human Blood or Human Plasma

Annex 15-Qualification and validation (into operation since 1 October 2015)

Annex 16-Certification by a Qualified Person and Batch Release

Annex 17-Parametric Release

Annex 19-Reference and Retention Samples

FDA

FDA publishes regulations and guidance documents for industry in the Federal Register. FDA's website also contains links to the CGMP regulations and guidance documents various resources to help drug companies comply with the law. FDA also conducts onsite audits and public outreach through presentations at national and international meetings and conferences on the subject of CGMP requirements.

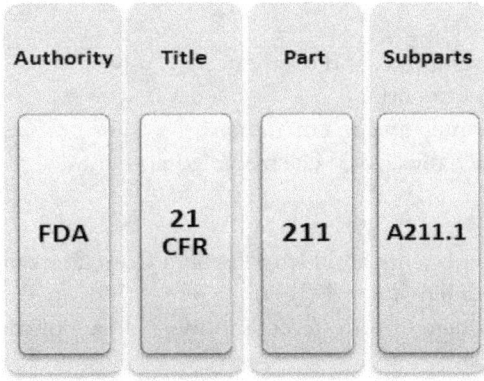

Figure 2: The FDA organises its regulations under titles. Within titles there parts and subparts.

Pharmaceutical quality affects every American. FDA regulates the quality of pharmaceuticals very carefully. The main regulatory standard for ensuring pharmaceutical quality is the Current Good Manufacturing Practice (CGMPs) regulation for human pharmaceuticals. Consumers expect that each batch of medicines they take will meet quality standards so that they will be safe and effective. Most people, however, are not aware of CGMPs, or how FDA assures that drug manufacturing processes meet these basic objectives. Recently, FDA has announced a number of regulatory actions taken against drug manufacturers based on the lack of CGMPs. This paper discusses some facts that may be helpful in understanding how CGMPs establish the foundation for drug product quality.

PART 211   Current Good Manufacturing Practice For Finished Pharmaceuticals

Subpart A--General Provisions
 § 211.1 - Scope.
 § 211.3 - Definitions.

Subpart B--Organization and Personnel
 § 211.22 - Responsibilities of quality control unit.
 § 211.25 - Personnel qualifications.
 § 211.28 - Personnel responsibilities.
 § 211.34 - Consultants.

Subpart C--Buildings and Facilities
 § 211.42 - Design and construction features.
 § 211.44 - Lighting.
 § 211.46 - Ventilation, air filtration, air heating and cooling.
 § 211.48 - Plumbing.
 § 211.50 - Sewage and refuse.
 § 211.52 - Washing and toilet facilities.
 § 211.56 - Sanitation.
 § 211.58 - Maintenance.

Subpart D--Equipment
 § 211.63 - Equipment design, size, and location.
 § 211.65 - Equipment construction.
 § 211.67 - Equipment cleaning and maintenance.
 § 211.68 - Automatic, mechanical, and electronic equipment.
 § 211.72 - Filters.

Subpart E--Control of Components and Drug Product Containers and Closures
 § 211.80 - General requirements.
 § 211.82 - Receipt and storage of untested components, drug product containers, and closures.
 § 211.84 - Testing and approval or rejection of components, drug product containers, and closures.
 § 211.86 - Use of approved components, drug product containers, and closures.
 § 211.87 - Retesting of approved components, drug product containers, and closures.
 § 211.89 - Rejected components, drug product containers, and closures.

World Health Organisation GMP Guideline Annexes

The WHO Essential Medicines and Health Products (EMP) Department works with countries to promote affordable access to quality, safe and effective medicines, vaccines, diagnostics and other medical devices. As part of this effort, the WHO publishes a number of guidance annexes that describe best practice quality requirements for specific areas within the life science industry.

List of WHO GMP annexes:

- WHO good manufacturing practices for pharmaceutical products: main principles
  Annex 2, WHO Technical Report Series 986, 2014
- Active pharmaceutical ingredients (bulk drug substances)
  Annex 2, WHO Technical Report Series 957, 2010
- Active pharmaceutical ingredients - bulk drug substances: Additional clarifications and explanations
- Pharmaceutical excipients
  Annex 5, WHO Technical Report Series 885, 1999
- WHO good manufacturing practices for sterile pharmaceutical products
  Annex 6, WHO Technical Report Series 961, 2011
- WHO good manufacturing practices for biological products
  Annex 3, WHO Technical Report Series 996, 2016
- WHO good manufacturing practices for blood establishments (jointly with the Expert Committee on Biological Standardization)
  Annex 4, WHO Technical Report Series 961, 2011
- Pharmaceutical products containing hazardous substances
  Annex 3 WHO Technical Report Series 957, 2010
- Investigational pharmaceutical products for clinical trials in humans
  Annex 7, WHO Technical Report Series 863, 1996
- Herbal medicinal products
  Annex 3, WHO Technical Report Series 937, 2006
- Radiopharmaceutical products
  Annex 3, WHO Technical Report Series 908, 2003
- Water for pharmaceutical use
  Annex 2, WHO Technical Report Series 970, 2012
- WHO guidelines on good manufacturing practices for heating, ventilation and air-conditioning systems for non-sterile pharmaceutical dosage forms
  Annex 5, WHO Technical Report Series 961, 2011
- Validation
  Annex 4, WHO Technical Report Series 937, 2006
- Guidelines on good manufacturing practices: validation, Appendix 7: non-sterile process validation
  Annex 3, WHO Technical Report Series 992, 2015

International Council for Harmonisation, ICH, GMP Guide

The International Council for Harmonisation of (Technical Requirements) for Pharmaceuticals for Human Use (ICH) brings together the regulatory authorities and pharmaceutical industry to discuss scientific and technical aspects of drug registration. Since its inception in 1990, ICH has gradually evolved, to respond to the increasingly global face of drug development.

- ICH Q7 Good Manufacturing Practice Guide for Active Pharmaceutical Ingredients

CHAPTER 2

Quality Management

What Is Quality?

Quality can be defined as the ability to consistently produce products meeting the same specifications time after time. Products must be safe, pure, uniform and effective. Specifications can be set down internally within a company, however, depending on the product, external specifications from regulators or standards may be required.

Patient safety is the primary focus of any pharmaceutical drug or medical device. This is the expectation of any patient or user. Secondly, the patient or user is interested in receiving an effective product. It is product specifications that ensure these criteria are accounted for.

What Is A Quality Management System?

A Quality Management System, often abbreviated to (QMS) is any system based on a collection of business processes that are primarily focused on providing safe and quality products that consistently meet customer requirements. The core themes of a QMS are outlined below.

Customer and Regulatory Focus

An understanding of the customer needs and requirements should be evident within the organisation and with the future vision of the company. The company should have an understanding of the regulatory landscape as this is subject to change over time. In turn the company should be positioned to respond to that change.

Leadership

To truly lead, one must be accepted in the hearts and minds of those they lead. Authentic leadership pays off. A leader should foster a sense of togetherness and common vision. A leader is anyone who influences or directs people either formally or informally. We are all leaders to some extent.

Involvement

Engagement by everyone across an organisation is now recognised as being key in the successful deployment of any Quality Management System. Everyone should have a voice within the company. As the saying goes "we are only as strong as the weakest link" is very apt.
The Process Approach

Systems Management

This essentially means that systems are defined and described in writing along with the appropriate responses to expected issues that arise. Effective systems management must ensure that the various systems work in support of each other and communicate effectively with one another.

Decision Making

In order to make the right decision, the person empowered to make the decision must be informed. To be correctly informed one must have the correct details and facts available. In a manufacturing environment the facts are essentially data and the analysis of data. During manufacturing or processing, data is generated as a result of monitoring and measurement of products and the related processes.

Supplier Management

Don't ruffle your suppliers' feathers. Security of supply is key in delivering products to customers or patients again and again, Raw materials or sub-components sourced from external suppliers must always be sourced at the right price and time with the emphasis on getting the best quality possible.

Continuous Improvement

For IS0 13485 continuous improvement refers to improving the effectiveness of the Quality Management System. It is harder to drive improvement of the product due to regulatory and practical requirements.

The key elements of a QMS are listed below.

Quality Policy: A company will document their commitment and approach to quality within their organisation. It usually sets out how they plan to achieve a high and consistent standard of quality. It should in some way speak to the customer or end user.

Quality Objectives: Quality objectives can be documented in a Quality Plan at site or organisational level. An effective way of defining quality objectives is use of the SMART method. SMART stands for Specific, Measurable, Achievable, Realistic and Timely.

Quality Manual: An in-house guidance document to provide a framework for achieving the quality objectives.

Organisational Structure and Responsibilities: Organisational charts can be used to map out the company structure. Roles and responsibilities can be documented in site quality plans, job descriptions and Standard Operating Procedures.

Data Management: A coherent approach to the provision, storage and maintenance of data.

Processes: Processes are defined and documented.

Resources: Resources must be properly understood, allocated and linked across the organisation.

Product Quality & Customer Satisfaction: The proper management and investigation of complaints is important to reduce future instances from reoccurring. Continual engagement with the end user or customer is critical.

Continuous improvement including corrective and preventive action- where continuous improvement projects and initiatives are encouraged and supported. The application of a CAPA system to ensure quality is maintained and consistent.

Maintenance: A Preventative Maintenance schedule is in place and managed accordingly.

Sustainability: All work practices are sustainable and consistent throughout the lifecycle of processes and products.

Auditing: Systems are auditable and maintained to allow internal or external review and audit.

Engineering Change Control: Where changes are required to validated processes or equipment, changes are managed and introduced under change control.

A common acronym used to highlight the aims of Good Manufacturing Practices (GMP) is SPUE which stands for Safe-Pure-Uniform-Effective. This definition is particularly suited to pharmaceutical products as the chemicals and drugs used need to be pure and free of contaminants. Furthermore, they need to be uniform, meaning they will have the same constituents from tablet to tablet and batch to batch. A description of each word is shown below:

SAFE- the product has the right ingredients if it is a drug product. It is packaged as intended and correctly labelled in order to provide identification and safe use.

PURE- it is free of contaminants, foreign matter, chemicals and harmful microbes.

UNIFORM- The product is manufactured consistently and will have the same quality between batches manufactured on different days.

EFFECTIVE- Ultimately, the product must be effective in treating the medical condition. To be effective, it requires the correct ingredients, the correct amount of ingredients and correct packaging to maintain the product stability over time.

The basic concepts of Quality Management, Good Manufacturing Practice and Quality Risk Management are inter-related. They are described here in order to emphasise their relationships and their fundamental importance to the production and control of medicinal products.

Figure 3

A Pharmaceutical Quality System must be appropriate for the manufacture of medicinal Products and incorporate management controls that ensure patient safety.

1) Product realisation: product realisation is the process of identifying market opportunities and user needs and bringing them forward through a design, planning stage that will result in a new product. A systematic approach to introducing new products is important to ensure consistency.

2) Product and process knowledge must be managed throughout all stages of a product from design and development to ultimate retirement. Training, formal documentation and design specifications all contribute to the knowledge pool.

3) Medicinal products are designed and developed in a way that takes account of the requirements of Good Manufacturing Practice;

4) Both production and control operations are clearly specified and Good Manufacturing Practice adopted

5) Managerial responsibilities are clearly specified

6) Arrangements are made for the manufacture, supply and use of the correct starting and packaging materials, the selection and monitoring of suppliers and for verifying that each delivery is from the approved supply chain. (e.g. Supplier approval processes)

7) Systems of control are established and maintained by developing and using effective monitoring and control systems for both process performance and product quality. Metrics help ensure processes are in control and product quality is been maintained

8) Continual improvement is facilitated through the implementation of quality improvements appropriate to the current level of process and product knowledge

9) Arrangements are in place for the prospective evaluation of planned changes and their approval prior to implementation taking into account regulatory notification and approval where required

10) Formal tools such as root cause analysis are applied during the investigation of deviations, suspected product defects and other quality issues. This can be determined using Quality Risk Management principles. In cases

Medicinal products should not be sold or supplied before a Qualified Person has certified that each production batch has been produced and controlled in accordance with the requirements of the competent authority and any other regulations relevant to the production, control and release of medicinal products. There should also be a process for self-inspection and/or quality audits that examines regularly the effectiveness and applicability of the Pharmaceutical Quality System.

Quality Control

Quality Control is that part of Good Manufacturing Practice which is concerned with sampling, specifications and testing, and with the organisation, documentation and release procedures which ensure that the necessary and relevant tests are actually carried out and that materials are not released for use, nor products released for sale or supply, until their quality has been judged to be satisfactory. The basic requirements of Quality Control are that:

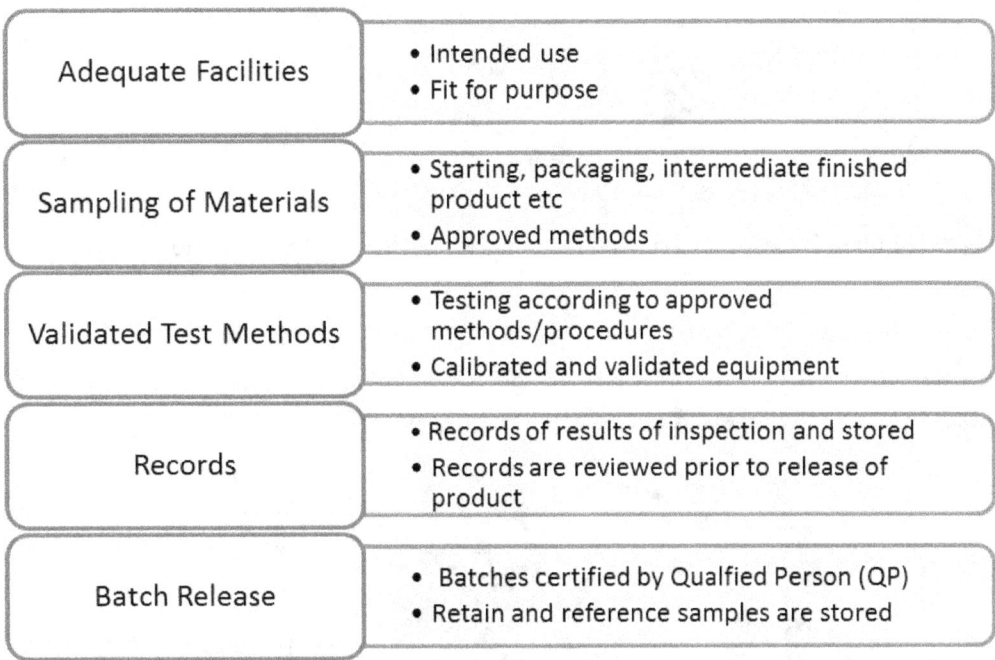

Figure 4: Control elements within a Quality Management System

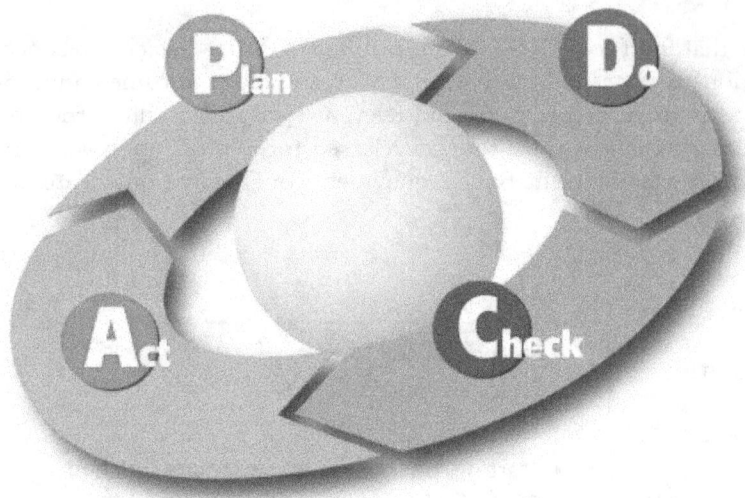

Figure 5: Plan, Do, Check Act, PDCA- continuous improvement methodology.

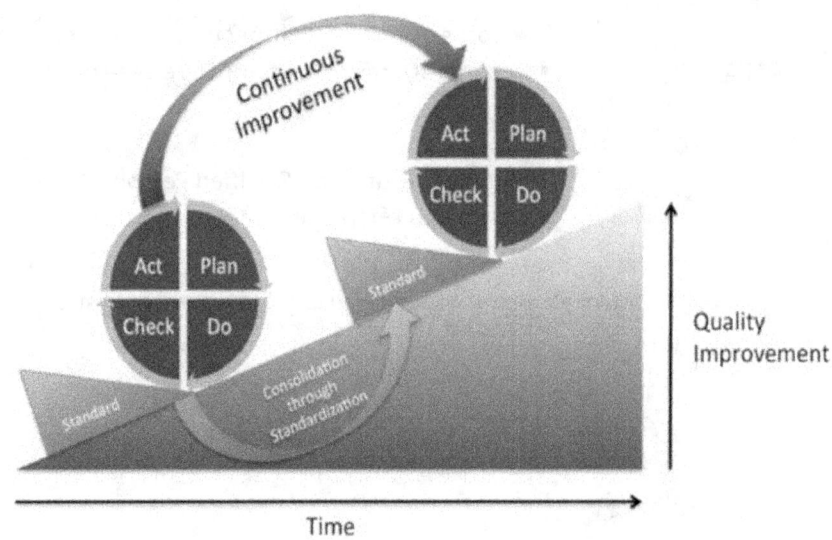

Figure 6: The PDCA implemented continuously over time

On-going Stability

After marketing, the stability of the medicinal product should be monitored according to a continuous appropriate programme that will permit the detection of any stability issue (e.g. changes in levels of impurities or dissolution profile) associated with the formulation in the marketed package.

The purpose of the on-going stability programme is to monitor the product over its shelf life and to determine that the product remains, and can be expected to remain, within specifications under the labelled storage conditions.

The protocol for an on-going stability programme should extend to the end of the shelf life period and should include at a minimum, the following:

(i) Number of batch(es) per strength and different batch sizes, if applicable
(ii) physical, chemical, microbiological and biological test methods
(iii) Acceptance criteria

(v) Description of the container closure system(s) or packaging
(vi) Testing intervals (time points)
(vii) Description of the conditions of storage

A summary of all the data generated, including any interim conclusions on the programme, should be written and maintained. This summary should be subjected to periodic review.

CHAPTER 3

Personnel

Introduction

Personnel are central to the application of CGMP and compliance to regulations. A every level throughout an organisation, people interact with materials, equipment and processes in order to deliver products to the market and patient. Personnel must therefore be suitably qualified and equipped to carry out their responsibilities effectively.

Table 1: GMP Personnel Requirements Overview

Overview of Personnel GMP Requirements					
Item	**PICS/s**	**Eudralex**	**FDA**	**WHO**	**ICH**
Reference	GMP Guide Intro/ Part 1	V4 Part 1, Chapter 2	21 CFR Part 210 Subpart B--Organization and Personnel	Annex 2	Q7 ICH
Key Headings	Principle General Key Personnel Training Personnel Hygiene Consultants	Key Personnel Personnel Hygiene Consultants	Responsibilities of quality control unit. Personnel qualifications. Personnel responsibilities. Consultants	Personnel Personnel Hygiene	Personnel Qualifications Personnel Hygiene Consultants
Key Words/ Themes	Training Authorised person Head of Quality Control Detailed hygiene programs Consultant records (Name, address, qualifications, type of service)	Key Personnel Qualified Person Training Duties of Quality Control Detailed hygiene programs Protective Garments	Training in GMP. Adequate number of personnel. Good health habits. Consultant records (Name, address, qualifications, type of service)	Key Personnel with heads of departments Authorized Persons Direct contact avoided Personnel Hygiene Procedures	Appropriate Education Training Responsibilities defined Good health habits. Suitable Clothing Smoking eating etc. restricted to designated areas.

Provisions in guidance and regulations are therefore made for Personnel in a Quality Management System. Despite advances in automation and computerised systems, people are centrally involved in day to day decisions. For this reason there must be sufficient and suitability qualified personnel to carry out all the tasks. Individual responsibilities should be clearly defined and understood by the persons concerned and recorded formally in procedures and job descriptions.

General

It may be an obvious point; however, manufacturers must ensure an adequate number of personnel with the necessary qualifications and practical experience are resourced to manufacturing. Having a broad base of people with the experience, knowledge and skills reduces the risk of quality issues. Responsibilities placed on any one individual should not be so extensive as to present any risk to quality.

Personnel should have specific duties recorded in written descriptions and adequate authority to carry out its responsibilities. Its duties may be delegated to designated deputies with a satisfactory level of qualifications.

Personal Hygiene

All personnel should be trained in the practices of personal hygiene. A high level of personal hygiene should be observed by all those concerned with manufacturing processes. Personnel should be instructed to wash their hands before entering production areas. Signage should be in place along with hand washing facilities. Hand washing demonstrations and training should be provided by a suitably qualified QC analyst or Microbiologist. Any person experiencing an illness or exhibiting open lesions or wounds that may adversely affect the quality of products should not be allowed to handle starting materials, packaging materials, in-process materials or medicines until the condition is no longer a risk to quality or patient safety.

Direct contact should be avoided between the operator's hands and starting materials, primary packaging materials and intermediate or bulk product.

CHAPTER 4

Buildings and facilities

Introduction

Facilities and utilities qualifications are typically pre-requisites to the validation of manufacturing equipment and systems. Much of the activity that deals with establishing a facility or building that is *fit for purpose* is managed under the broad heading of commissioning and qualification (C&Q). The terms C&Q are often used interchangeably and in practice some overlap in activity is expected. Commissioning can be defined as the planned, documented, and managed engineering approach to the start-up and handover of facilities, systems, and equipment to the end-user. It must deliver a safe and functional environment that meets the pre-defined design and user requirements.

In strict terms, qualification is more concerned with the confirmation and documentation showing that equipment or systems are properly installed and functional. Qualification forms part of validation, but the individual qualification steps do not equal a validated process. The establishment of a user requirements specification (URS) and detailed design specifications ensure that the building or facility will meet end users' needs and that it is fit for the intended purpose.

It also provides a level of protection to the contracting company responsible for the project or facility construction. Post-URS approval requires an approved Design Qualification (DQ). This provides verification and a documented record that the proposed design is suitable for the intended purpose. Further verification including IQ/OP/PQ should be applied as required based on the system impact and criticality of facilities/utilities.

Risk and Impact Assessment

A risk-based qualification process should assess the potential of a system to impact the product quality. The boundaries of any system (HVAC, compressed air supply etc.) should be identified in order to help establish the scope of any system and determine if it has a direct, indirect or no impact on product quality.

Direct Impact: a system that can directly impact product quality.

Indirect Impact: where a system is not expected to directly impact the product quality but supports or is ancillary to a direct impact system.

No Impact: a system that does not directly impact product quality and does not support a direct impact system.

For a HVAC System supplying a classified area, only once the air enters the room must the air quality meet the classified designation. The Critical Quality Attributes (CQAs) are routinely monitored through the Environmental Monitoring Program and the Critical Process Parameters (CPPs) should be monitored through the calibrated and validated Environmental Monitoring System (QBMS). The Direct Impact (level 1) for the HVAC systems are indicated on the boundary diagram shown below.

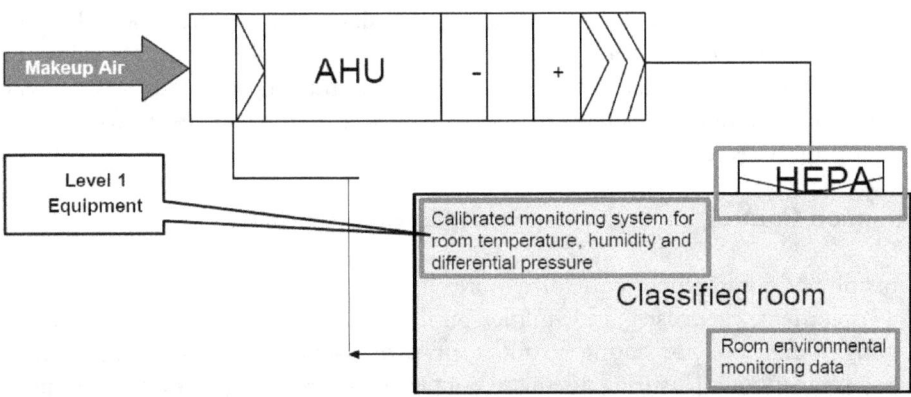

Figure 7: HVAC System Boundary Diagram (Level 1). Each individual system is represented by a green box. Separate qualifications should be performed for each one. The room environmental monitoring system is typically called a Building Management System (BMS). The *calibrated* monitoring system for room temperature, humidity and differential pressure is called a QBMS. Where the "Q" stands for quality indicating the system is used to monitor critical parameters.

Qualification Levels

Qualification levels are often used within companies to classify the criticality of equipment or systems. Level 1 requires the highest level of verification.

Level 1: a system where an **undetected change** in system performance poses a significant risk to the product and product safety. Level 1 systems require the highest degree of qualification and validation. This should include URS/DQ/IQ/ OQ/ PQ.

Level 2: a system where a change that **may be detected** in system performance poses a significant risk to product and product safety. These systems require a level of qualification including IQ, however OQ and PQ testing may not be required. This should be based on the intended use of the system, impact on product quality and overall risk.

Level 3: All other systems.

Typically IQ or equivalent testing is sufficient. **Note:** other requirements or qualifications should be based on risk.

The level of qualification and validation testing required for any system should be based on a risk assessment, examining the criticality of the system and environment. Risk assessments should consider the following points:

- Building design and construction features
- System boundaries and complexity
- Potential product impact
- Environmental controls and monitoring systems
- Potential impact to operator safety
- Type of qualification/validation (e.g. prospective, concurrent, or retrospective)

Controlled-not-classified (CNC) environments, utilities, and facility control systems also require adequate qualification/validation. Again, the impact on product quality should be determined in order to shape any validation. Routine monitoring test locations as well as alert and action levels should be determined in advance of any validation for environmental monitoring or utility systems.

Contamination Control

The philosophy of containment control requires it to be applied across all inputs that make up a facility- equipment, processes, and utilities and so on. Containment is primarily concerned with keeping things in- preventing product or processing agents from egressing into the surrounding atmosphere. Ensuring adequate containment protects personnel who interact with the process, equipment and systems. Aseptic processing often deals with biological agents or compounds that may be harmful to operators or technicians. A secondary concern of containment is protection of the environment. Containment also compliments efforts in contamination prevention. As with Aseptic processing the risk to the patient and product must be at the forefront of activity. Risk based approaches and tools should be used to identify potential risks and put in place adequate controls and mitigations. Any assessment should take into account all the following systems:

> Facility layout
> Drainage Systems
> HVAC requirements
> Location and adequacy of utilities
> Personnel flow and procedures for entering and leaving
> Behavioral requirements of personnel in the clean room
> Flow of materials and products to prevent cross-contamination and mix-ups between products and between dirty and clean or sterile and non-sterile equipment and products
> Design to avoid cross-contamination when manufacturing live biological agents, e.g. local exhaust air HEPA filtration, dedicated air handling units.

Material flow

The design and layout of any manufacturing area should facilitate the effective flow of materials. This is a fundamental requirement no matter what the industry, e.g. medical devices, pharmaceuticals, bio pharmaceuticals and even non-regulated engineering companies that assemble, machine or fabricate products. However, with the manufacture of medicinal products that are required to be sterile imposes a greater level of control and thought. With regard to Aseptic processing facilities material flows do not only require efficient and effective flow of materials, the activity should support the requirements of Aseptic processing while minimizing any risk of contamination. Identifying critical processing zones is step in ensuring the right building design and controls are implemented. Isolators and Aseptic filling require the highest classification with strict environmental controls. Secondary packaging operations such as cartoning are often completed in areas controlled and operated to a lower classification.

Design and layout of facilities should:

- ➢ Maintain microbiological integrity of the identified critical processing zones
- ➢ Prevent or minimise contamination from outside critical processing zones
- ➢ Control the flow of materials by restricting access to trained and authorized personnel

Material Transfer

Material transfer from the outside of cleanrooms to the inside is completed via material air locks or hatches. Material air locks and hatches ensure that there is clear separation between controlled clean areas and less clean areas. Many suppliers provide products that are double bagged. This provides an added level of control when transferring materials. The outer bag can be removed within the air lock thus providing a clean inner product. Material air locks also allow the sanitization of products. Tools and other items must be clean and dirt free.

Controls that prevent personnel from the clean area and less clean area been present in the material air lock at the same time. This can be achieved by training and educating staff on the importance of contamination control. A simple visual check of the air lock to confirm it is vacant can be done in order to avoid mixing of personal from different zones. Decontamination procedures are necessary to ensure materials or tools entering the controlled area are decontaminated.

Material Air lock considerations:
- ➢ Interlocked doors
- ➢ Access control
- ➢ Sanitation/ Cleaning procedure
- ➢ Double or triple bagged products
- ➢ Dedicated trolley for air locks

Disinfection and Cleaning Agents

When materials are been transferred via an air lock, consideration must be given to the status of materials and products. As a rule, no cardboard or unnecessary paper should enter a cleanroom. Wooden pallets are not acceptable as they can carry dirt and microorganisms and wood cannot be sanitized due to its porous nature. Soft fabric cases often used to carry tools should also be avoided as the material can carry dirt and grease. Cleaning and disinfecting agents should be tested and approved prior to their use onsite. The choice of agents should be backed up with studies that demonstrate the effectiveness of disinfectants and cleaning agents.

Gown up Areas

Gowning rooms are designed in order to minimize contamination and facilitate the orderly change over from street clothes to scrubs and/or gowns. Hand washing facilities help reduce the risk of humans carrying unwanted microorganisms into the aseptic processing area. The design of the room should result in clear separation between the less clean side and the clean side. This can be achieved with a step over segregating the two areas.
Other features of gowning rooms should include:

- ➢ Storage lockers for street clothes
- ➢ Gown and Garment storage
- ➢ Body Length mirrors
- ➢ Hand Washing /Drying and disinfection facilities

GMP Zoning

Selecting a suitable classification for a room or manufacturing facility depends on several factors. Firstly, it can be said that sterile products require a more stringent set of criteria than non-sterile products. However, there is an extensive range of products and medical devices that are sterile but are used in different ways and consist of different materials and technology. Some sterile products are single use only and used for short term purposes and then disposed of. Other sterile products are used subcutaneously for longer periods or even require implantation. Therefore, the design of a facility along with its HVAC specification must be appropriate to the product being manufactured. High risk products require greater control. The goal of facilities and HVAC systems is to minimise contamination and the associated risks. Using a "sterile versus non-sterile" rule of thumb is not adequate when classifying a room or facility. Standards including EN ISO 14644-1 and guidelines such as EU cGMP Guidelines EudraLex volume 4 Annex 1 (2008) should be consulted in order to fully understand the requirements of each ISO classification and grade of room.

ISO classifications do not specify room occupancy states but when a designation is applied, the occupancy state must be stated in the relevant documentation or procedure. The most relevant European Guideline (Annex 1 of the EU cGMP Guideline) lists four classification grades and their associated particulate limits in the 'at rest' and 'in operation' conditions. In general, for the sterile and non-sterile products, similar classes are applied, but in non-sterile production the producer could assign their classes, having similar particulate concentration, temperature, pressure etc. but lower air-change rate could be used.

Types of Contamination:

- cross contamination (of a product/material with another product/material)
- non-microbial particulate contamination (non-viable particles)
- biological/microbiological contamination (viable particles/micro-organisms)

Factors Influencing Contamination Cleanliness Levels in the Manufacturing Processes:

- process
- air cleanliness
- personnel hygiene and clothing
- work practices
- material design (material of construction, surface finishes, room finishes, equipment, open system/enclosed system
- utensils, etc.)
- material cleanliness

<u>Room Air Classification (By Limits of Particulate Contamination)</u>

ISO CLASS	FDA	cCMP	Permissible particle number in 1 m3					
			0,1 µm	0,2 µm	0,3 µm	0,5 µm	1 µm	5 µm
1			10	2				
2			100	24	10	4		
3	1		1,000	237	102	35	8	
4	10		10,000	2,370	1,020	352	83	
5	100	A	100,000	23,700	10,200	3,520	832	29
6	1,000	B	1,000,000	237,000	102,000	35,200	8,320	293
7	10,000	C				352,000	83,200	2,930
8	100,000	D				3,520,000	832,000	29,300
9						35,200,000	8,320,000	293,000

Figure 4: Table showing ISO classes and EudraLex Grades A-D.

	Maximum permitted number of particles per m³ equal to or greater than the tabulated size			
	At rest		In operation	
Grade	0.5 µm	5.0µm	0.5 µm	5.0µm
A	3 520	20	3 520	20
B	3 520	29	352 000	2 900
C	352 000	2 900	3 520 000	29 000
D	3 520 000	29 000	Not defined	Not defined

Figure 5: maximum permitted airborne particle concentration for each grade. Showing both "at rest" and "in operation" conditions. (EU V4 Annex 1). The EU guidance given for the maximum permitted number of particles in the "at rest" column corresponds approximately to the ISO classifications.

Room Air Classification (By Limits of Microbial Contamination)

The HVAC systems help maintain the viable (microbial) limits within a specific area. These limits are defined in Annex 1 of the EU GMP Guide as shown below.

Grade	Recommended limits for microbial contamination (a)			
	air sample cfu/m³	settle plates (diameter 90 mm) cfu/4 hours (b)	contact plates (diameter 55 mm) cfu/plate	glove print 5 fingers cfu/glove
A	< 1	< 1	< 1	< 1
B	10	5	5	5
C	100	50	25	-
D	200	100	50	-

Figure 6: Recommended limits for microbial contamination

Environmental Grade A (Aseptic)

Grade A is reserved for critical processes in manufacturing sterile products, product components or product contact. This is generally achieved using isolator technology which maintains a barrier to the background environment or surrounding room.

Grade A Operations include:

➢ Aseptic processing of sterile ingredients
➢ Filling of sterile products not for terminal sterilisation
➢ Stopper insertion
➢ Crimp Capping

Environmental Grade B

Grade B is used for supportive work for aseptic processing corresponds to ISO 14644 (Part 1) Class 5 ("at rest") and Class 7 (when "in operation"). Grade B areas typically serve as the background environment of Grade A areas for aseptic processing.

Environmental Grade C

Suitable for non-critical processing steps, Grade C corresponds to ISO 14644 Part 1 Class 7 ("at rest") and Class 8 ("in operation"). Grade C operations include:

> ➢ Clean side of material air locks and gowning rooms
> ➢ Filling of products that are to be terminally sterilised

Environmental Grade D

Grade D at least corresponds to ISO 14644 Part 1 Class 8 ("at rest" / no definition for "in operation").

> ➢ Clean section of material air locks and final compartments of gowning rooms
> ➢ Dispensing of raw materials and excipients and preparation of solutions for sterile products to be sterile filtered and terminally sterilised
> ➢ Background environment for transfer and crimp capping of stoppered containers with sterile products

Compliance Tests for GMP Zones

Test	Requirements
Particle count test	Test covers verification of cleanliness. Dust particle counts to be carried out and result printed. The number of readings and positions of tests should be defined in accordance with ISO 14644-1 Annex B5
Air pressure difference	This test is used to verify non cross-contamination. Log of pressure differential readings to be produced or critical plants should be logged daily, preferably continuously. A 15 Pa pressure differential between different zones is recommended. Refer to ISO 14644-3 Annex B5
Airflow volume	To verify air change rates. Airflow readings for supply air and return air grilles to be measured and air change rates to be calculated. Refer to ISO 14644-3 Annex B13
Airflow velocity	To verify unidirectional flow or containment conditions. Air velocities for containment systems and unidirectional flow protection systems to be measured. Refer to ISO 14644-3 Annex

	B4
Filter leakage tests	To verify filter integrity. Filter penetration tests to be carried out by a competent person to demonstrate filter media, filter seal and filter frame integrity. Only required on HEPA filters. Refer to ISO 14644-3 Annex B6
Containment leakage	To verify absence of cross-contamination. Demonstrate that contaminant is maintained within a room by means of: • airflow direction smoke tests • room air pressures. Refer to ISO 14644-3 Annex B4
Recovery	To verify clean-up time. Test to establish time that a cleanroom takes to recover from a contaminated condition to the specified cleanroom condition. Should not take more than 15 minutes. Refer to ISO 14644-3 Annex B13
Airflow visualisation	To verify required airflow patterns. Tests to demonstrate air flows: • from clean to dirty areas • do not cause cross-contamination • uniformly from unidirectional airflow units Demonstrated by actual or video-taped smoke tests. Refer to ISO 14644-3 Annex B7

Clean Room Design Considerations

Air handling unit (AHU) -Air Intake Quality

Seasonal Variations

All locations on earth except latitudes near the equator experience seasonal temperature changes. The changes are a consequence of the earth's orbital motion about the sun, coupled with the tilt of earth's axis of rotation with respect to its orbital plane. Design criteria should be based on published temperature data. The HVAC system design should consider the following:

Standard Operating Conditions: These are climatic conditions against which the systems must be designed to operate, control, and maintain required conditions. (These may be based on published data, which are only exceeded 2.5% or 1% of the time).

Extreme Operating Conditions: These are climatic conditions against which the systems must be designed to operate, without manual intervention, and without damage to the systems or the facility. Based on product / process risk assessments, extreme or standard conditions shall be used for HVAC design for dedicated areas.

Location

Based on the building layout, foot-print and design intent, a suitable and adequate space must be identified for HVAC location. This must include provision of chilled water, heating systems, ducts and drainage. HVAC plants must be accommodated in designated HVAC plant rooms or interstitial areas.

Air Intake

During the design phase, the air intake locations should be selected that ensure air is in the best environmental condition. The below considerations help to achieve a strong starting point:

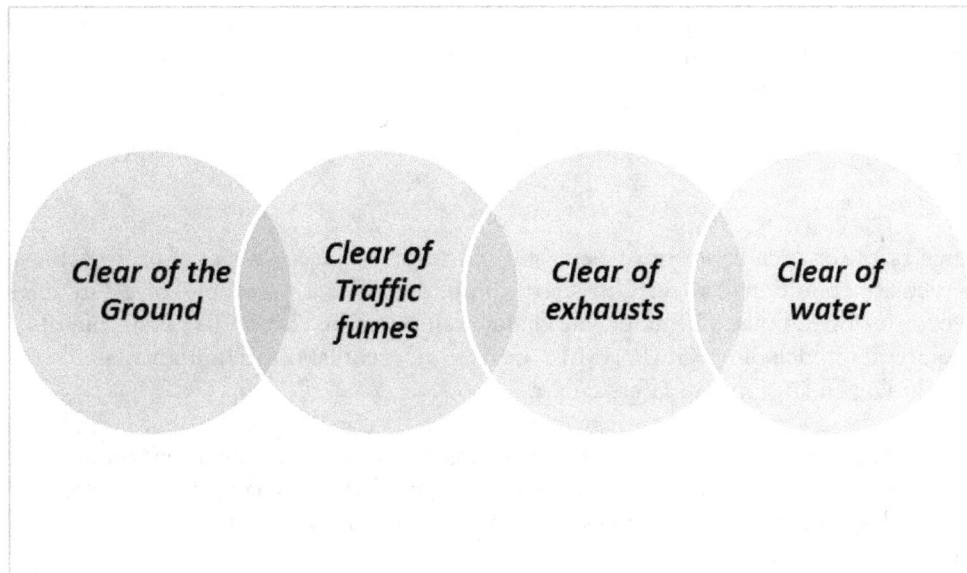

Figure 8: Air intake considerations

Thermal Load

Thermal load can be defined as the amount of heat energy to be removed from an inner environment by equipment (HVAC) used to maintain that environment at the design temperature when worst case external temperature(s) are being experienced. The thermal load requirement should be calculated for the following:

- ➢ Max summer conditions
- ➢ Minimum winter conditions
- ➢ High rainfall
- ➢ Standard operation
- ➢ Extreme operating conditions

Room Recovery Time

Room recovery time to return to the required pressure differential and cleanliness Specification should be minimized.

Dust, Vapour, or Fume Control

Highlight areas requiring dust, vapour, gas and/or fume control on the room data sheet. These areas must be controlled to remove the possibility of product contamination and to ensure the safety of the operator and environment. Areas requiring 100% fresh air or extraction to atmosphere may require greater airflow or other measures within the room to maintain environmental conditions.

In order to meet the appropriate level of cleanliness, HVAC systems require sufficient filtration to provide "clean" air to prevent contamination of the product. Pre-filters and main filters are normally suitable for most operations; however, HEPA filters are required to prevent particulate or microbial contamination for higher-classification areas

Air Change Rates

The air change rates for each room must be calculated to be sufficient for cleanup to achieve specified particulate conditions "at rest" in static conditions after a maximum of 20 minutes from completion of operations. The actual air change rate must be chosen to satisfy the most stringent requirement including GMP, GLP, heat gain, ventilation requirements and/or occupancy, including an appropriate safety factor.

The air change rate must be optimized for energy savings; however, specific attention must be paid to air locks where a greater air change rate must be applied. Air changes can be reduced (e.g., setback modes) in some circumstances ("at rest" mode, with no production activity and no personnel interventions).

Room Environmental Conditions

Other environmental conditions to be controlled, such as temperature and relative humidity, depend on the product and nature of the operations carried out in those areas. These parameters should not interfere with the defined cleanliness standard.

Temperature Requirement

The normal operating temperature requirement for each classification .Temperature and humidity must be appropriate to the product and process Consideration should be made for specific product and process requirements.

The normal operating humidity requirement

Particulate Levels

Particulate levels are specifically defined for each room classification "at rest" and "in operation". The levels are controlled though air filtration, facility design, gowning requirements, and decontamination

Room Exhaust

Where there is a risk of active compounds being present in extracted air, filters should be fitted, preferably at the room, to prevent contamination of ductwork and the environment. The filters must be selected based on the particle size distribution of the products to be handled.

HVAC system design

The HVAC system must be appropriately selected using the specific design requirements as outlined above. The system must be able to provide clean, conditioned air to the specified areas to meet all of the quality requirements. The most important precursor to HVAC design is the comprehensive definition of the function and performance required followed by the selection of an appropriate system. A poor selection can lead to unnecessarily high-energy consumption, and operational deficiencies. HVAC systems can be divided into two main types:

All-air systems rely on the movement of large quantities of air through a central air handling unit to control room conditions, as well as provide for ventilation requirements.

They have the advantage of being relatively simple with most of the unit situated in one location; however, they are very space consuming. All-air systems tend to be relatively inflexible and not ideal for areas that are likely to need environmental alteration on a regular basis.

These HVAC systems are used for areas that have a lot of small zones, each with slightly different thermal loads but which requires constant ventilation. These systems can have poor energy efficiency if a lot of reheat is required. These are typically used in large manufacturing areas, and laboratories with many small rooms.

Dust Extraction and Collection

It is essential to capture dust as close as possible to the point of generation without affecting the process. In most cases dust capture should be within 100 mm from the point of release. Air velocity is the key parameter in dust capture.

Pharmaceutical and chemical applications have specific collection requirements as any dust build-up in the system is likely to be of a pharmacologically active nature, sensitizing, toxic and/or corrosive. It is vital to maintain transport velocities and minimizes any potential for cross contamination.

A typical system should have a minimum transport velocity of 18 m/s, but this may need to be higher if heavy particles are to be collected. This velocity must be maintained throughout the system to prevent dust from dropping out in the ducts.

The dust collection must be configured with the hazardous nature of the dust in mind. A clearly defined disposal procedure for the collected dust (e.g., bag-in / bag-out system for filter and dust bin) needs to be understood at the design stage. HVAC unit shall meet EN 1886 and EN 13053 requirements.

Fans

Certified performance curves are required to verify correct fan operation. Fans that may be subjected to high temperatures, humidity, corrosive fumes or other hazardous atmospheres should be constructed using non-reactive, non-corrosive, suitable and approved materials (such as epoxy painting). Whenever H_2O_2 or other disinfection application is planned, material compatibility certificates shall be supplied by the vendor.

Fans must be selected to supply the design volume, taking into account the assumption that filters are half clogged, except for the terminal filter which shall be considered to be fully clogged according to EN 13053. If the terminal filter is HEPA, clogging shall be considered according to EN 1822 and the target volume is 80% of the given maximum clogged specified value.

Filtration

Face-fitting filters shall be used in all cases, as slide-in filter elements never give a good seal. The installation must be such that the airflow pushes the filter against the seal. The face velocity across the filter section shall not exceed 2 m/s. For ventilation and air conditioning applications, two minimum filtration stages are required. For certain applications, return air filtration will be required to contain highly active materials (e.g., viruses or potent compounds). Normally, these filters should be changed from the room side. However, since those filters must be integrity tested, it is recommended to place one filter in the main return duct before the exhaust fan and design return duct network, in order to ensure tightness of the duct between the room and the filter (bag-in / bag-out filter change systems should be provided for BSL-3 areas).In case of live biological agent biocontainment, decontamination up to the filter must be proven. The grade of filter and technical solution must be selected based on the product particle size distribution and occupational exposure band (OEB) level.

For most HVAC applications, dehumidification is best achieved by the use of cooling coils. It should be noted that dehumidification is a very high consumer of energy and should only be used if there is a real process need. When areas are not in use, the dehumidifier should be turned off, if possible.

When room humidity must be maintained below 50% during warm weather, an absorption dryer may be necessary unless the room temperature can be increased within specification to compensate.

Normal practice is to optimize size and efficiency of the absorption dryer by first removing as much moisture from the air as possible by cooling. The design of absorption dryers is normally based on a slowly rotating desiccant wheel.

Air is passed through the wheel and dried by the desiccant coating (guidance: lithium chloride especially if the wheel is not used frequently and silica gel if used permanently and with low humidity target). It is not normally necessary to size a dryer to handle the entire air volume. Drying a proportion of air and re-mixing to achieve the desired moisture content is usually sufficient.

Air humidification may be necessary during cold weather when introducing fresh air to spaces that require humidity control. When air humidification is necessary, humidifiers should be selected on the following basis:

> ➢ direct steam injection using steam
> ➢ direct steam injection using self-generative electric or gas steam humidifier.

Clean steam is required, However when industrial steam is used a quality gap assessment should be conducted with Quality team taking into account the regulatory requirements 21 CFR 173.310 for operator safety and to avoid product contamination and mainly boiler feed water treatment must be free of volatile additives such as amines and hydrazines.

Humidifiers should be located before the fan and the final filter which will remove any particulate generated. At least 300 mm clearance should be allowed upstream and 1 m downstream between humidifier manifolds and coils, attenuators etc. (general recommendation to be confirmed through calculation note provided by the vendor). A single manifold or multiple manifolds in parallel may be used to meet the humidification requirements as per manufacturer's recommendations.

Sound Attenuators

Sound attenuators should be provided as necessary, to achieve the specified noise levels within occupied spaces and to minimize external noise nuisance assessment can confirm the necessity to use acoustic media (enveloped in polyester film), that is inert and corrosion-resistant at normal operating conditions. Material quality shall be equivalent to that specified for HVAC unit or ducts.
Sound attenuators should be installed in the air handling unit or ductwork. The use of sound attenuators in the air supply and air return should be based on requirements for fresh air inlet and air exhaust, and according to external noise levels that might need to be maintained at or below the ambient site noise levels.

Dampers

The provision of sufficient dampers is essential for proper control. To minimize noise transmission into the room, these should be mounted as far as possible from the diffuser.

Carefully evaluate the space-by-space pressure control that will be used in the design. Static pressure control via hard balance or dynamic control via air terminal control units are both appropriate. Consideration should be given to the overall project size, the complexity of the facility and the project budget.

Automatic volume controllers are recommended for regulating air volume independently of supply pressure. They can be selected for constant volume, variable volume or dualduct mixing applications. Automatic low-leakage fresh air and exhaust air shutoff dampers are strongly recommended to isolate the HVAC network. Fresh air dampers shall be Class 3 minimum (maximum leakage preventing coil freezing). Whenever fumigation is performed shutoff damper shall ensure Class 4 leakage rate. Where dampers are required to provide modulating control of airflow, they must be selected to provide an appropriate level of control authority. This will normally mean a damper smaller than the duct size.

Heating and cooling

Heating mode: Low pressure hot water (LPHW) is the preferred heating medium for HVAC applications and should be used whenever practicable. Electrical heating should be avoided due to fire risk and should limited to low power coil and in locations where no other energies are available. Hazard operability analysis (HAZOP) must be conducted if electrical heating is being considered. Cooling mode: Chilled water is the preferred cooling medium for HVAC applications and should be used whenever practicable.

The direct expansion of refrigerant in coils is an acceptable method of cooling, particularly on small isolated plants, or where lower temperatures are needed for dehumidification or for cold room. This system, however, does not normally give close control. Direct expansion coils should only be used with extreme care on variable air volume systems (if speed driver available on compressors).

Heating Coils

The face velocity of air across heating coils should not exceed 2 m/s. Coils should be made of material suitable for applicable constraints. Drains shall be located outside the casing of the HVAC unit. Coils shall be removable.

Cooling coils

Cooling coils have been identified as potential sources of microbial contamination; therefore, careful design is required to prevent water carryover and to ensure that drain pans do not retain water. Double tube, non-welded units are recommended. The face velocity of air across cooling coils should not exceed 2 m/s. Where necessary, stainless steel or plastic eliminator blades should be provided to prevent any moisture carryover. Where provided, these must be removable for cleaning.

Ductwork

For most applications galvanized steel ductwork will be the most appropriate form of construction; however, stainless steel or plastic construction may be necessary where there is a higher risk of corrosion due to moisture or fumes (exhaust ducts usually). Where operating pressures above 2,000 Pa are necessary, fully welded construction is recommended. For contained ducts (e.g., exhaust duct before bag-in / bag-out filter), air tightness Class C shall be followed (EN 12237). For BSL-3, fully welded construction should be considered.

Generally ductwork should be constructed to an appropriate local standard, suitable for the maximum design pressure (positive or negative), such as those published by Sheet metal and Air Conditioning Contractors' National Association (SMACNA) in the USA, Building and Engineering Services Association (B&ES) in the UK
Where flexible connections are proposed these must be designed for the same pressure as the ductwork. Solid ducted connections are preferred for final connections to terminal HEPA filter housings. For applications where flexible connections to diffusers are used, these should be no longer than 500 mm and nominally straight.

Special consideration must be given to fume extract ducts where these pass through fire barriers. Using fire dampers should be avoided where the loss of extraction could make a fire situation worse. An alternative design, such as the use of fire-rated ductwork, may be necessary in these cases. A thorough risk assessment must be conducted.

Simple Representation of HVAC system

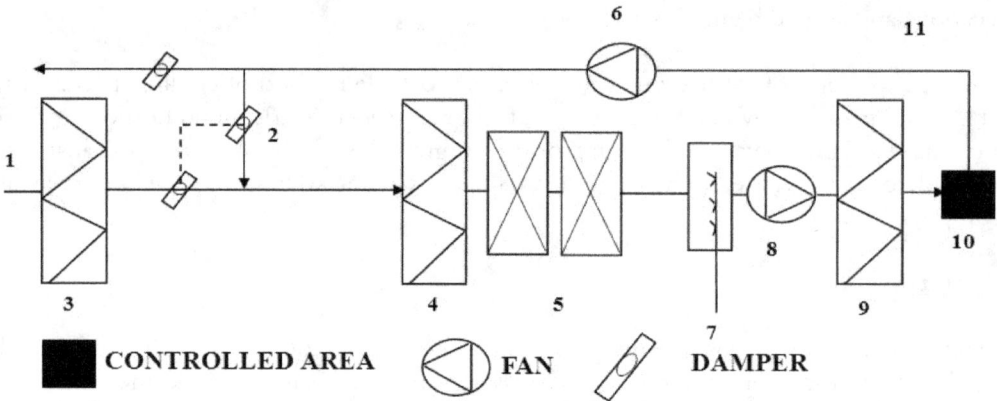

Figure 9: Simple HVAC diagram

Position	Description
1	Fresh Air Intake (°C, %RH, Flow rate)
2	Dampers
3	Filter creating a Differential Pressure
4	Filter creating a Differential Pressure
5	Control Valves for cooling fluid
6	Exhaust Fan
7	Steam Flow Rate
8	Supply Fan
9	Filter creating a Differential Pressure
10	Controlled room/ area
11	Extraction

Parameter	Description
Temperature	The HVAC must be capable of operating over a range of temperatures and accurate to a tolerance. Temperature probes/detectors must be placed at various points to provide feedback and control.
Relative Humidity	Relative humidity must be monitored continuously. Typically, humidity sensors should be effective over an operational range (e.g. 5-95% R.H.) Accuracy should also be no less than ±3%
Air flow/ Air Pressure	Air flow is proportional to the square root of the differential pressure.
Dampers	Dampers are used to control inlet and outlet airflow.
Valves	Ball valves or globe valves control the flow of air. Valves are designed with specific safety features to meet the intended use.(e.g. CLOSED without energy supply-cooling valve, OPEN without energy supply-heating valve.

Environmental Monitoring

An environmental monitoring program is required for GMP controlled areas. The purpose of such programs is to document, define and describe parameters to be monitored, monitoring frequency and methods. Environmental monitoring is a regulatory requirement. It also demonstrates that the GMP areas are been controlled and are fit for purpose.

Key Requirements of Environmental Monitoring

Identification and classification of environmental areas that require monitoring

Test methods and sampling procedures

Defined testing frequencies

Sample locations based on Risk

Microbial monitoring of personnel

Monitoring of non viable particles

Monitoring of temperature, relative humidity and differential pressures

Defined alert and action levels for each environmental area

Trending of Enviromental data

Change Control

Figure 10 : Key requirements of Environmental Monitoring

Other parameters such as those controlled by the HVAC system (air changes/hour etc.) should also be verified according to a defined schedule.

Grade A, B and C

> ➤ Viable and non-viable particles monitored under operational conditions
> ➤ Risk-based approach to sampling points and represent high risk/ critical positions

<u>Grade D</u>

- ➤ Non-viable particles must be measured at-rest conditions
- ➤ Viable particles measured under operational conditions

Building Management Systems

A Building Management System (BMS) is an automated control system that is used to manage a building and facilities heating, ventilation and air conditioning, security, fire protection systems and so on. It is made up of many different Input / Output subsystems, controller(s), server(s) and workstation(s) communicating over a control network to control, monitor, alarm and trend equipment. . BMS systems are also referred to as a Facilities Management/Monitoring System (FMS), Energy Management System (EMS), Building Automation System (BAS) or other equivalent.

Environmental Monitoring System (EMS) are automated control systems consisting of Input / Output subsystems, controller(s), server(s) and workstation(s) communicating over a control network to monitor, alarm and trend environmental critical process parameters such as temperature, humidity, differential pressure, conductivity, cooler / refrigerator status amongst others. Suggested classification of BMS based on intended use:

Building Management System (BMS)	
System Classification	GxP
Data Usage	Data not used for GxP impacting decisions. Engineering use only
Monitoring	No Critical process parameters are monitored by the system
Controls	No GxP equipment
System Boundaries	Up to the point of use of the system or equipment
Validation	Not required

Environmental Monitoring System (BMS)	
System Classification	Non GxP
Data Usage	Data may be used to make quality decisions and product release decisions. Data is used to determine compliance.
Monitoring	Critical process parameters are monitored by the system
Controls	Critical alarm limits are controlled
System Boundaries	From the point of use
Validation	Not required

CHAPTER 5

Process Equipment

General Requirements

Process equipment must be located, designed, constructed and maintained to suit the intended use and manufacturing operations. The layout and design of equipment must aim to minimize the risk of errors and permit effective cleaning and maintenance. Both equipment and facilities should minimise cross-contamination.

While the various annexes and guides (E.g. WHO, ICH, etc.) arguably the FDA provides more specific requirements and details.

Intended Purpose

The intended purpose of equipment should be documented in a User requirements specification.

Lab Equipment

Laboratory equipment and instruments should be suited to the testing procedures and give the required accuracy according to acceptance specifications.

Washing, cleaning and drying equipment should not act as a source of contamination.

Contamination Control

Production equipment should not present any hazard to the products. The parts of the production equipment that come into contact with the product must not be reactive, additive, or absorptive to an extent that would affect the quality of the product.

Suitable material should also be non-shedding. High grade stainless steels and acetal plastics are often the material of choice when it comes to equipment, contact surfaces and utensils.

Closed equipment should be used whenever appropriate and practical. Non-dedicated should have at validated cleaning protocol in place and documented procedures that detail the cleaning and sanitisation requirements between the production of different pharmaceutical products.

Maintenance

Effective maintenance procedures and practices falls under the heading of Equipment. Maintenance often covers cleaning activities required in order to insure products are not contaminated. With reference to FDA 21 CFR, the following requirements are set out.

"Written procedures shall be established and followed for cleaning and maintenance of equipment, including utensils, used in the manufacture, processing, packing, or holding of a drug product. These procedures shall include, but are not necessarily limited to, the following:

(1) Assignment of responsibility for cleaning and maintaining equipment;
(2) Maintenance and cleaning schedules, including, where appropriate, sanitizing schedules;
(3) A description in sufficient detail of the methods, equipment, and materials used in cleaning and maintenance operations, and the methods of disassembling and reassembling equipment as necessary to assure proper cleaning and maintenance;
(4) Removal or obliteration of previous batch identification;
(5) Protection of clean equipment from contamination prior to use;
(6) Inspection of equipment for cleanliness immediately before use.
(c) Records shall be kept of maintenance, cleaning, sanitizing, and inspection"

Calibration

ICH, 07 provides the greatest amount of detail when compared to other Annexes and regulations.

➤ *"Control, weighing, measuring, monitoring and test equipment that is critical for assuring the quality of intermediates or APIs should be calibrated according to written procedures and an established schedule."*

➤ *Equipment calibrations should be performed using standards traceable to certified standards, if existing*

➤ *Records of these calibrations should be maintained.*

➤ *The current calibration status of critical equipment should be known and verifiable.*

➤ *Instruments that do not meet calibration criteria should not be used.*

➤ *Deviations from approved standards of calibration on critical instruments should be investigated to determine if these could have had an impact on the quality of the intermediate(s) or API(s) manufactured using this equipment since the last successful calibration."*

Computerised Systems

GMP impacting computerized systems need to be qualified and formally validated. Typically a GxP assessment determines the GAMP classification of a system and thus determines the scope and depth of validation.

Commercially available and qualified software usually does not require the same level of testing as bespoke or customised software. Data integrity of computer systems must be demonstrated in any validation activities.

Table 2: GMP Process Equipment GMP Requirements

Overview of Process Equipment GMP Requirements					
Item	**PICS/s**	**Eudralex**	**FDA**	**WHO**	**ICH**
Reference	GMP for Medical Products Part I, Chapter 3	EU GMP V4 Part 1, Chapter 3	CFR - Code of Federal Regulations Title 21, Part 211	Annex 2, Section 3.0	ICH, Q7, Process Equipment
Key Headings	Equipment 3.34-3.44 Note: Content as per Eudralex	Equipment 3.34-3.44	§ 211.63 - Equipment design, size, and location. § 211.65 - Equipment construction. § 211.67 - Equipment cleaning and maintenance. § 211.68 - Automatic, mechanical, and electronic equipment. § 211.72 - Filters.	Equipment	5.1 Design and Construction 5.2 Equipment Maintenance and Cleaning 5.3 Calibration 5.4 Computerized Systems
Key Words/ Themes	Note: Content as per Eudralex	Intended purpose of Equipment Minimum risk of contamination Maintenance Range and precision	Material and surface requirements Contamination prevention Data backup of Computerised systems Filters-non shedding, max pore size of 0.2 microns	Location, design and installation Cleaning Regime Non dedicated equipment Closed equipment Note: Some identical requirements to Eudralex	Intended use Surfaces Qualified operating range Lubricants, coolants etc. Closed or contained equipment Current drawings

CHAPTER 6

Documentation and Records

Introduction

Good documentation is an essential part of the quality assurance system and is relevant across all departments and functions within a manufacturing company. Controlled documents define the specifications and procedures for all materials and methods of manufacture and control strategies.

General Requirements

➢ Documents should be designed, prepared, reviewed, approved and distributed in accordance with approved processes.

➢ Documents should comply with relevant parts of the manufacturing and marketing authorizations.

➢ Documents should be approved, signed and dated by the appropriate responsible persons. No document should be changed without authorization and approval.

➢ Documents should have unambiguous contents: the title, nature and purpose should be clearly stated. They should be laid out in an orderly fashion and be easy to check.

➢ Documents must not allow any error to be introduced through the reproduction process.

Good Documentation Practices

This section provides an easy-to-understand guide to the subject of Good Documentation Practices. Good Documentation Practices (commonly abbreviated to GDP or GDocP) is a term used to describe standards by which documents are created, modified and maintained. The need for GDP is driven by the general requirement of GMP (Good Manufacturing Practices)

GDP is a practical skill that is required within the life science sector (medical device, pharmaceutical and so on). It can be broadly divided into two streams; GDocP practices and how they apply to electronic document and secondly, GDocP for handwritten entries including initial and dating and recording of data and test results by hand. GDocP is fundamental in achieving compliance to Good Manufacturing Practices (GMP). It is required in the U.S. by the FDA's Code of Federal Regulations and in Europe by the governing body EudraLex. If GDocP is not practiced it jeopardises the integrity of data and written records. This can lead to the falsification of data which is a serious regulatory offence. Admittedly, implementing and maintaining GDP takes time, effort and resources, however, there are some benefits that come with it. Most importantly, Good Documentation Practice is an expected practice in regulated industry as trust and ethics are fundamental to business.

It is important to maintain accurate, complete, up-to-date and consistent information to ensure patient safety and reduce any potential risk to patients. Practicing GDP equally helps to reduce observations raised on inadequate documentation practices at times of audit by regulated bodies such as the FDA. It helps to improve communication and efficiency within companies. If GDP is not followed it can call into question other processes and procedures within a company.

Documentation Creation

The principles of GDP should be applied at the document creation stage. As most people are familiar with softcopy or electronic documents, some of these points are obvious but nonetheless need to be made. All documents should be electronically written and not handwritten except for execution of protocols, test results and adding entries. Documents that are approved controlled should be:

Accurate and free from errors
Have revision or version controlled
Should have an effective date or date of release

Approval of Documents

Document approval must be completed by trained and appropriately experienced personnel. Often companies will use an approval matrix which explains which people are required to approve each document. For example, an EHS (Environment Health and Safety) officer would be required to approve a risk assessment.

Signatures

A signature on any document is legally binding so remember to read and understand what is being signed for. Every signature should also include the date in the correct format. If a signature appears within the same document alongside initials, substituting a full signature with initials and date is generally acceptable. This practice is common when large documents are being completed.

Date and Time Format

A standardised approach to dates and times is important especially within large global organisations. For instance, in the USA, the norm is to place the month before the date, whereas in Ireland and Great Britain it is common to write the day of the month followed by the month. Most companies would define their date and time format in an SOP or procedure. The date and time format can also be configured in Word documents and Excel worksheets to align with a companies preferred date and time format.

Handwritten Entries

When a handwritten entry is required such as a signature or a test result, indelible ink must be used. Many companies will have an SOP or procedure that states the specific ink colour required. If an entry of a test result or test data isn't completed at the time of execution, this constitutes a late entry. Backdating an entry or signature is forbidden. Always use the correct and current date.

How Are Mistakes Corrected?

This is a critical area of GDP. Failure to follow the requirements of GDP when correcting mistakes is the most common failure when it comes to documentation in industry. The method of correcting mistakes using GDP allows for a person looking at the document for the first time to clearly see the original entry and the corrected entry. This maintains the integrity of the document. In order to identify the changes and corrections, certain rules must be followed. No overwriting is allowed and white-out or Tipp-Ex is not allowed. The correct way to make any changes or corrections by hand is shown in the diagram below.

Accuracy

Accuracy of information provided in documents is critical in the life science industry. As the end user is a patient, inaccurate records or documents could cause serious injury or death. Controlled documents are also legal documents and could be called upon if recalls, litigation or investigations arise.
Many documents used in the manufacture of medical devices are designed to record information or test results. These test results are then used to disposition (pass or fail) batches of product. Inaccurate information could risk the release and distribution of defective product. This has a potential impact on both the business and the patient or user.

Blank Spaces or Blank Fields

On completion of a document such as a logbook or record, no blanks spaces should be left unfilled. This is to avoid late entries and also to prevent confusion. Blank spaces or blank fields should have a diagonal line drawn neatly across the space, the letters "N/A" written and the entry signed and dated. If the reason for "N/A" is not evident then it is wise to include an explanatory note or sentence.

Data Transcription

Transcribing is the process of transferring data from one source to another. This is often required when raw data is involved. When data is in raw format it may need to be entered into a Microsoft Excel sheet. When transcribing data remember that all original raw data must be stored in case it is needed at a future date. After the data is transcribed it must be verified by a second person to check for any errors or omissions.

Revision Control

Controlled documents should always have a version number or revision number electronically on each page of the document. This is similar to books which always list what edition they are e.g. first edition or second edition. Revision control is a key element of the Quality Management Systems in place in regulated industries. As the need for changes in the document arises, the controlled document can be amended/updated. With each update the version number revises also. Some companies will use alphabetic revision control and to a lesser extent numeric revision control (Version A, Version B or Version 01, Version 02).

Management of Attachments

Attachments to controlled documents can include training records, data sheets, lab results and so on. It is important that attachments are identified for traceability purposes. If the attached becomes detached from the main document, the attachment should be identifiable. It is best practice to include a reference number on the attachment if available. If the attachment consists of several pages, each page should be numbered in Page X of Y format if not electronically done so. And remember, hand written entries must be accompanied by a signature and date. Always use staples to attach documents together. Glue or paper clips are not acceptable.

Management of Documents through Their Lifecycle

GDP applies to all the different stages of a document's lifecycle. These stages include creation, review, approval, issuing, completion of records, revision, updating, retirement and storage. Storage a.k.a. retention is an important stage and often a legal requirement for medical devices and pharmaceutical products. For consumer OTC medicines a 5-year retention of quality records often suffices. For implants such as TKRs or Total Knee Replacements, a 90-year retention period is required. This ensures that traceability and a quality record is available if the need arises.

Test Results

This section provides an overview on the correct handling of test results. Test results can be generated from various types of product testing such as visual inspection, dimensional inspection and chemical analysis. The recording of all test results should be completed on an approved form. This is to ensure that the correct information is being recorded and the same approach is taken by different people who might have to complete testing.

Calculations

There are different ways calculations can be completed. Many simple calculations can be done by an individual using a calculator, alternatively, a software package such as Minitab or an Excel sheet can be used to complete complex calculations. It should be clear to the reader what calculation is required, what the formula is and how the calculation is completed.
If the formula used is not included on the sheet, it should be referenced in a controlled document. Care is also required when recording numbers of several decimal places in length, as rounding error can be introduced.

Units of Measurement

The most important thing to remember is consistency in units of measurement when recording data or making calculations. Consult your company procedure if available to determine the correct units of measurement. Many U.S. companies use imperial units e.g. inches, pounds etc. In Europe the International System of Units or SI is used, e.g. millimetres and kilograms.

Batch Records

Batch records document critical information relating to the manufacture of products. Depending on the product, it can include dispensed weights of raw materials. It may also include critical parameters, times and dates of critical steps, in process test results and so on.

- ➢ Batch records should be reviewed and checked for:
- ➢ Accuracy
- ➢ Legibility
- ➢ Correct document version
- ➢ Completeness
- ➢ Correct references to supporting documents
- ➢ a unique batch or identification number
- ➢ be dated
- ➢ signed when issued/approved

With reference to ICH Q7, the following requirements are specified in a clear and concise format beneficial to the manufacturer.

"Documentation of completion of each significant step in the batch production records (batch production and control records) should include: − Dates and, when appropriate, times;

− Identity of major equipment (e.g., reactors, driers, mills, etc.) used;

− Specific identification of each batch, including weights, measures, and batch numbers of raw materials, intermediates, or any reprocessed materials used during manufacturing; − Actual results recorded for critical process parameters;

− Any sampling performed;

− Signatures of the persons performing and directly supervising or checking each critical step in the operation;

− In-process and laboratory test results;

− Actual yield at appropriate phases or times;

− Description of packaging and label for intermediate or API;

− Representative label of API or intermediate if made commercially available;

− Any deviation noted, its evaluation, investigation conducted (if appropriate) or reference to that investigation if stored separately;

− Results of release testing."

Laboratory Records

Laboratory control records should include complete data derived from all tests conducted to ensure compliance with established specifications and standards, including examinations and assays.

A description of samples received for testing, including the material name or source, batch number or other distinctive code, date sample was taken, and, where appropriate, the quantity and date the sample was received for testing; ICH Q7 states the following requirements.

— *A statement of or reference to each test method used;*

— *A statement of the weight or measure of sample used for each test as described by the method; data on or cross-reference to the preparation and testing of reference standards, reagents and standard solutions;*

— *A complete record of all raw data generated during each test, in addition to graphs, charts, and spectra from laboratory instrumentation, properly identified to show the specific material and batch tested;*

— *A record of all calculations performed in connection with the test, including, for example, units of measure, conversion factors, and equivalency factors;*

— *A statement of the test results and how they compare with established acceptance criteria;* —

The signature of the person who performed each test and the date(s) the tests were performed; and

— *The date and signature of a second person showing that the original records have been reviewed for accuracy, completeness, and compliance with established standards.*

Complete records should also be maintained for:

— *Any modifications to an established analytical method;*

— *Periodic calibration of laboratory instruments, apparatus, gauges, and recording devices;*

— *All stability testing performed on APIs; and*

— *Out-of-specification (OOS) investigations.*

(Ref: ICH, Q7.)

CHAPTER 7

Materials Management

Introduction

The key theme of effective materials management is control of materials from incoming stage through the manufacturing process. Specifications and testing support the control of materials ensuring they are meeting the key quality requirements to allow consistent manufacturing and quality products.

Table 3: GMP Materials Management

Overview of GMP Materials Management					
Item	**PICS/s**	**Eudralex**	**FDA**	**WHO**	**ICH**
Reference	GMP for Medical Products Part I, Chapter 5 Production	EU GMP V4 Part 1, Chapter 5 Production 5.27	CFR - Code of Federal Regulations Title 21, Part 211	Annex 2, Section 14.0	ICH, Q7, Section 7, Materials management
Key Headings	Starting materials Processing operations-intermediate and bulk products Packaging materials Packaging operations Finished Products Rejected, recovered and returned materials	Starting Materials Processing operations: intermediate and bulk products Packaging materials Finished products Rejected, recovered and returned materials Product shortage due to manufacturin	Subpart E-- Control of Components and Drug Product Containers and Closures	General Starting materials Packaging materials Intermediate and bulk products Finished Products Rejected, recovered, reprocessed and reworked Recalled product Return goods	General controls Receipt and quarantine Sample and testing Storage Re-evaluation

		g constraints			
Key Words/ Themes	Specifications CofA	Active substances Excipients	Specifications Retesting Approved materials CofA	Specifications CofA	Specifications Sampling CofA Storage

Starting Materials

The designated name of the product and the internal code reference where applicable;

- ➢ Manufacturers batch number
- ➢ the status of the contents (e.g. quarantined, on test, released)
- ➢ the expiry date or a date beyond which retesting is necessary

Packaging Materials

The purchase, handling and control of primary and printed packaging materials should be as for starting materials. Particular attention should be paid to printed packaging materials. They should be stored in secure conditions so as to exclude the possibility of unauthorized access. Roll feed labels should be used wherever possible. Cut labels and other loose printed materials should be stored and transported in separate closed containers so as to avoid mix ups. Packaging materials should be issued for use only by designated personnel following an approved and documented procedure.

Intermediate

Intermediate products can be simply described as raw materials that may have been mixed and processed to some degree or other. Intermediate and bulk products should be kept under appropriate conditions and must be used within specified dates and according to specifications.

Finished Product

Finished products should be held in quarantine until their final release, after which they should be stored as usable stock under conditions established by the manufacturer.

For the approval and maintenance of suppliers of active substances and excipients, the following is required:

Active substances

Supply chain traceability should be established and the associated risks, from active substance starting materials to the finished medicinal product, should be formally assessed and periodically verified. Appropriate measures should be put in place to reduce risks to the quality of the active substance.

The supply chain and traceability records for each active substance (including active substance starting materials) should be available and be retained by the EEA based manufacturer or importer of the medicinal product.

Audits should be carried out at the manufacturers and distributors of active substances to confirm that they comply with the relevant good manufacturing practice and good distribution practice requirements. The holder of the manufacturing authorisation shall verify such compliance either by himself or through an entity acting on his behalf under a contract. For veterinary medicinal products, audits should be conducted based on risk.

Further audits should be undertaken at intervals defined by the quality risk management process to ensure the maintenance of standards and continued use of the approved supply chain.

Excipients

Excipients and excipient suppliers should be controlled appropriately based on the results of a formalised quality risk assessment in accordance with the European Commission 'Guidelines on the formalised risk assessment for ascertaining the appropriate Good Manufacturing Practice for excipients of medicinal products for human use'.

For each delivery of starting material the containers should be checked for integrity of package, including tamper evident seal where relevant, and for correspondence between the delivery note, the purchase order, the supplier's labels and approved manufacturer and supplier information maintained by the medicinal product manufacturer. The receiving checks on each delivery should be documented.

Prevention of Cross-contamination

Cross-contamination should be prevented by attention to design of the premises and equipment. This should be supported by attention to process design and implementation of any relevant technical or organizational measures, including effective and reproducible cleaning processes to control risk of cross-contamination.

A Quality Risk Management process, which includes a potency and toxicological evaluation, should be used to assess and control the cross-contamination risks presented by the products manufactured. Factors including; facility/equipment design and use, personnel and material flow, microbiological controls, physico-chemical characteristics of the active substance, process characteristics, cleaning processes and analytical capabilities relative to the relevant limits established from the evaluation of the products should also be taken into account. The outcome of the Quality Risk Management process should be the basis for determining the necessity for and extent to which premises and equipment should be dedicated to a particular product or product family. This may include dedicating specific product contact parts or dedication of the entire manufacturing facility.

Suggested Technical Measures

- Dedicated manufacturing facility (premises and equipment)
- Self-contained production areas having separate processing equipment and separate heating, ventilation and air-conditioning (HVAC) systems. It may also be desirable to isolate certain utilities from those used in other areas
- Design of manufacturing process, premises and equipment to minimize opportunities for cross-contamination during processing, maintenance and cleaning
- Use of "closed systems" for processing and material/product transfer between equipment
- Use of physical barrier systems, including isolators, as containment measures
- Controlled removal of dust close to source of the contaminant e.g. through localised extraction
- Dedication of equipment, dedication of product contact parts or dedication of selected parts which are harder to clean (e.g. filters), dedication of maintenance tools;
- Use of single use disposable technologies
- Use of equipment designed for ease of cleaning
- Appropriate use of air-locks and pressure cascade to confine potential airborne contaminant within a specified area
- Minimising the risk of contamination caused by recirculation or re-entry of untreated or insufficiently treated air
- Use of automatic clean in place systems of validated effectiveness
- For common general wash areas, separation of equipment washing, drying and storage areas.

Suggested Organisational Measures

- Dedicating the whole manufacturing facility or a self-contained production area on a campaign basis (dedicated by separation in time) followed by a cleaning process of validated effectiveness
- Keeping specific protective clothing inside areas where products with high risk of cross-contamination are processed
- Cleaning verification after each product campaign should be considered as a detectability tool to support effectiveness of the Quality Risk Management approach for products deemed to present higher risk
- Depending on the contamination risk, verification of cleaning of non product contact surfaces and monitoring of air within the manufacturing area and/or adjoining areas in order to demonstrate effectiveness of control measures against airborne contamination or contamination by mechanical transfer

- Specific measures for waste handling, contaminated rinsing water and soiled gowning;
- Recording of spills, accidental events or deviations from procedures
- Design of cleaning processes for premises and equipment such that the cleaning processes in themselves do not present a cross-contamination risk
- Design of detailed records for cleaning processes to assure completion of cleaning in accordance with approved procedures and use of cleaning status labels on equipment and manufacturing areas
- Use of common general wash areas on a campaign basis

CHAPTER 8

Rejection and Re-use of Materials

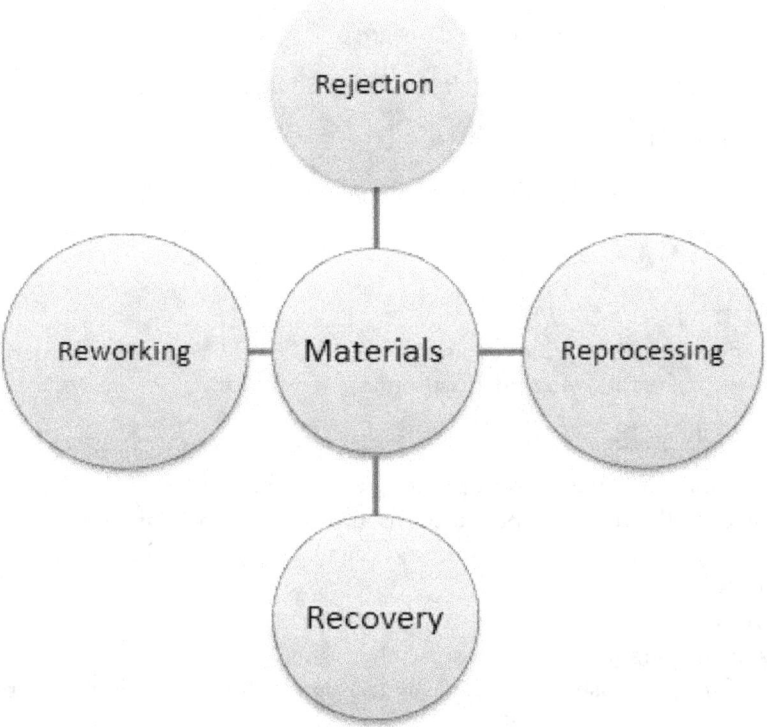

Figure 11: Material uses

Rejection

Intermediates and components failing to meet established specifications should be identified as such and quarantined according to a procedure. These items can be reprocessed or reworked as described below.

Reprocessing

Reprocessing by repeating a manufacturing step or a chemical or physical process of an established manufacturing process is generally considered acceptable.

Reprocessing should involve evaluation to ensure that the quality of product and must not adversely impact the safety of the finished product.

Recovery of Materials and Solvents

Recovery (e.g. from mother liquor or filtrates) of reactants, intermediates, or the API is considered acceptable, provided that approved procedures exist for the recovery and the recovered materials meet specifications suitable for their intended use.

Returns

Records of returned intermediates or APIs should be maintained. For each return, documentation should include:

- Name and address of the consignee

- Intermediate or API, batch number, and quantity returned

- Reason for return

- Use or disposal of the returned intermediate or API

Testing Of Materials

The tests performed should be recorded adequately. EU GMP V4 Part 1 Chapter 6: Quality Control recommends the following information as a minimum.

➢ Name of the material or product and, where applicable, dosage form
➢ Batch number and, where appropriate, the manufacturer and/or supplier
➢ References to the relevant specifications and testing procedures
➢ Test results, including observations and calculations, and reference to any certificates of analysis
➢ Dates of testing
➢ Initials of the persons who performed the testing
➢ Initials of the persons who verified the testing and the calculations, where appropriate

Sampling Checklist

The sample taking should be done and recorded in accordance with approved written procedures that describe:

➢ The method of sampling
➢ The equipment to be used
➢ . The amount of the sample to be taken
➢ Instructions for any required sub-division of the sample
➢ The type and condition of the sample container to be used
➢ The identification of containers sampled
➢ Any special precautions to be observed, especially with regard to the sampling of sterile

➢ or noxious materials
➢ The storage conditions
➢ Instructions for the cleaning and storage of sampling equipment

CHAPTER 13

Validation

Introduction

The term 'Validation Lifecycle' refers to the entire lifecycle, beginning with the initial requirements of a product or process and identifying CPPs and CQAs. The cycle continues through Commissioning and Qualification (C&Q), PQ and PV, requalification and ending with decommissioning or the end of life of a product line.

Process Validation is defined as "establishing documented evidence which provides a high degree of assurance that a specific process consistently produces a product meeting its predetermined specifications and quality attributes."

Why is Validation required?

There are several factors that require Validation activities within the life science industry. Above all, validation works to ensure patient safety and products that are fit for purpose and reliable time after time. However, regulations provide the legal incentive to validate processes within the medical device and medicinal industry. Validation activity also has secondary effect of fostering consistency in introducing new products and processes across different departments and sites.

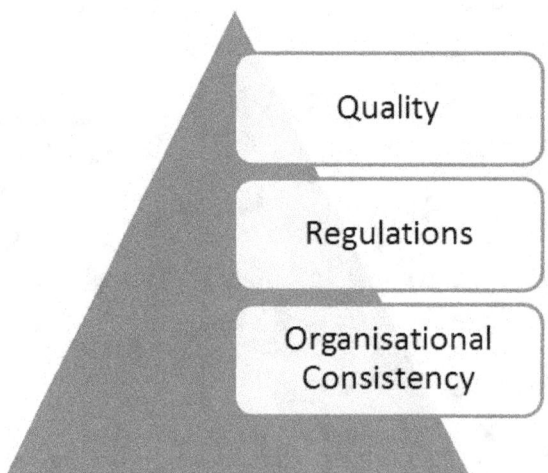

Figure 12 : Drivers of Validation

In many respects, regulations are the key driver as with any legal requirement, the manufacturer is required to fulfill their statutory obligations. Notified bodies such as the UK,

MHRA, US FDA specify rules and guidance in respect of pharmaceutical and medical products.

As the largest economy is the world, the United States is a leading user of regulated products. In turn, many countries throughout the world manufacture products with the intent to supply the US market. Therefore, manufacturers must meet the regulatory requirements set down by the US FDA. For Medical devices, 21 CFR Part 820 requires Validation to be completed for equipment and processes. For Pharmaceuticals, 21 CFR Part 211 also makes provision for Validation. In Europe, validation for the manufacture of medicinal products is a requirement of EU GMP V4, Medicinal product for human and veterinary use.

Summary of GMP Regulations and Standards

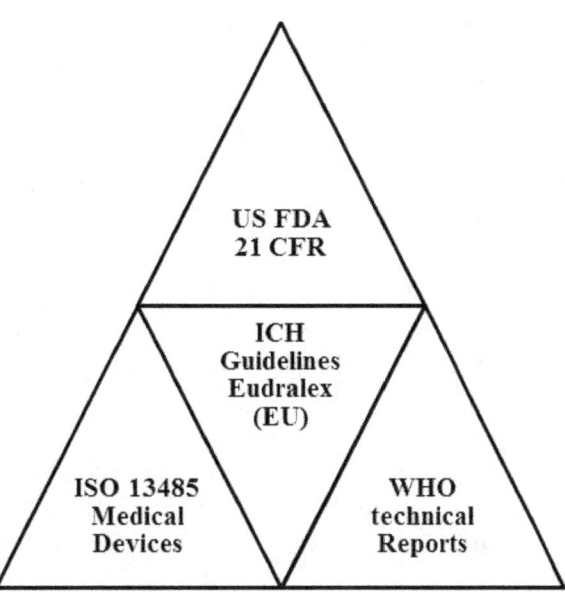

Figure 13: Key regulations and standards

Table 4: GMP Validation Requirements

Overview of Validation Requirements					
Item	PICS/s	Eudralex	FDA	WHO	ICH
Reference	GMP Part 1 PE 009-13 (Part I),Chapter 5, Production (Validation)	Eudralex V4 Part I, Chapter 5, Production	21 CFR Part 820.75	WHO, Annex 2	ICH, Q7 Section 12. Validation
Key Headings	Validation	Validation	Process Validation	Qualification and Validation	Validation Policy Validation documentation

Key Words/ Themes	Defined procedures Suitability Consistency Amendments to processes Periodic critical revalidation	Demonstrate suitability Periodic critical revalidation		Critical Aspects Validation Plan Direct/indirect impact Maintain validation status Ongoing programs of review	Qualification Approaches to Validation Process Validation Program Periodic Review
					Critical parameters/attributes DQ/IQ/OQ/ PQ Periodic review

PICS/s

The PIC/s guidance on Validation is limited to describing the basic requirements of a validation program such as having a validation procedure, qualifying a new process so that it is "suitable and "consistent".

PE 009-13 (Part I), Chapter 5, states:

➤ *"Validation studies should reinforce Good Manufacturing Practice and be conducted in accordance with defined procedures. Results and conclusions should be recorded.*

➤ *When any new manufacturing formula or method of preparation is adopted, steps should be taken to demonstrate its suitability for routine processing. The defined process, using the materials and equipment specified, should be shown to yield a product consistently of the required quality.*

➤ *Significant amendments to the manufacturing process, including any change in equipment or materials, which may affect product quality and/or the reproducibility of the process should be validated.*

➤ *Processes and procedures should undergo periodic critical revalidation to ensure that they remain capable of achieving the intended results."*

Eudralex

"Validation studies should reinforce Good Manufacturing Practice and be conducted in accordance with defined procedures. Results and conclusions should be recorded.

When any new manufacturing formula or method of preparation is adopted, steps should be taken to demonstrate its suitability for routine processing. The defined process, using the materials and equipment specified, should be shown to yield a product consistently of the required quality.

Significant amendments to the manufacturing process, including any change in equipment or materials, which may affect product quality and/or the reproducibility of the process, should be validated.

Processes and procedures should undergo periodic critical re-validation to ensure that they remain capable of achieving the intended results." (Ref: GMP V4 Part I, Chapter 5)

FDA

For Medical Devices, FDA 21 CFR Subpart G, Part 820 specifies the requirements for full verification or validation:

(a) Where the results of a process cannot be fully verified by subsequent inspection and test, the process shall be validated with a high degree of assurance and approved according to established procedures. The validation activities and results, including the date and signature of the individual(s) approving the validation and where appropriate the major equipment validated, shall be documented.

(b) Each manufacturer shall establish and maintain procedures for monitoring and control of process parameters for validated processes to ensure that the specified requirements continue to be met.

(1) Each manufacturer shall ensure that validated processes are performed by qualified individual(s).
(2) For validated processes, the monitoring and control methods and data, the date performed, and, where appropriate, the individual(s) performing the process or the major equipment used shall be documented.

(c) When changes or process deviations occur, the manufacturer shall review and evaluate the process and perform revalidation where appropriate. These activities shall be documented.

For medicinal/ pharmaceutical drug products 21 CFR 211.100(a) and 211.110(a) requires that drug products be produced with a high degree of assurance of meeting all the attributes they are intended to possess.

WHO

WHO GMP guidance requires that, each pharmaceutical company identifies what qualification and validation work is required to prove that the critical aspects of their particular operation are controlled. A validation plan should identify and describe what activities are required to be undertaken.

Annex 2 identifies the key requirements with regard to qualification and validation to ensure:

➢ the premises, supporting utilities, equipment and processes have been designed in accordance with the requirements for GMP (design qualification or DQ)

➢ the premises, supporting utilities and equipment have been built and installed in compliance with their design specifications (installation qualification or IQ)

➢ the premises, supporting utilities and equipment operate in accordance with their design specifications (operational qualification or OQ)

➢ a specific process will consistently produce a product meeting its predetermined specifications and quality attributes (process validation or PV, also called performance qualification or PQ).

ICH

ICH Q7 provides arguably provides the greatest amount of detail with regard to validation in a GMP environment.

Similar to other organisations, ICH requires a company to develop a "Validation Policy" to describe and document approaches to validations etc. ICH gives good guidance on validation in respect of Active pharmaceutical ingredients and the importance of critical process parameters and critical quality attributes. The approaches to Process Validation (Prospective, Concurrent etc.) also align with FDA requirements.

Key considerations for Validation of APIs

➢ Defining critical product attributes of APIs
➢ Identifying process parameters that could affect the critical quality attributes of API's
➢ Process Validation (PV should provide documented evidence that the process, operated within established parameters, can perform effectively and reproducibly to produce an intermediate or API meeting its predetermined specifications and quality attributes.

ICH Q7 key definitions regarding Qualification / Validation

"Design Qualification (DQ): documented verification that the proposed design of the facilities, equipment, or systems is suitable for the intended purpose.

Installation Qualification (IQ): documented verification that the equipment or systems, as installed or modified, comply with the approved design, the manufacturer's recommendations and/or user requirements.

Operational Qualification (OQ): documented verification that the equipment or systems, as installed or modified, perform as intended throughout the anticipated operating ranges.

Performance Qualification (PQ): documented verification that the equipment and ancillary systems, as connected together, can perform effectively and reproducibly based on the approved process method and specifications."

The Four Types of Process Validation

Process validation is a regulatory requirement of Good Manufacturing Practices (GMPs) for both pharmaceuticals (21CFR 211) and medical devices (21 CFR 820).

Prospective validation

Establishing documented evidence in advance of process implementation that a process or system operates as intended. This is the preferred approach and is most common when new products must be validated before commercial manufacturing.

Concurrent validation

Establishing documented evidence that a processes operates as intended, based on information generated during process implementation. Concurrent means that the outputs are performance of the system is monitored at the same time a manufacturing which can include commercial lots.

Retrospective validation

Retrospective validation is used for facilities or processes that have not completed formal Validation. Historical data or a retrospective review can provide the evidence that the process or facility is operated as intended. This type of validation is uncommon.

Revalidation

Revalidation involves the re-execution of validation activities in order to maintain a validated state. This can be a result of substantial changes to Product attributes or specification or changes to the manufacturing process itself. Other reasons a partial or full revalidation may be required involve instances where product quality issues have increased.

Stages of Process Validation

Process validation can be divided into in three stages:

Stage 1 – Process Design: The commercial manufacturing process is defined during this stage based on knowledge gained through development and scale-up activities.

Stage 2 – Process Qualification: During this stage, the process design is evaluated to determine if the process is capable of reproducible commercial manufacturing.

<u>Stage 3</u> – Continued Process Verification: Ongoing assurance is gained during routine production that the process remains in a state of control.

Before any batch from the process is commercially distributed for use by consumers, a manufacturer should have gained a high degree of assurance in the performance of the manufacturing process such that it will consistently produce APIs and drug products meeting those attributes relating to identity, strength, quality, purity, and potency. The assurance should be obtained from objective information and data from laboratory-, pilot-, and/or commercial scale studies. Information and data should demonstrate that the commercial manufacturing process is capable of consistently producing acceptable quality products within commercial manufacturing conditions. A successful validation program depends upon information and knowledge from product and process development. This knowledge and understanding is the basis for establishing an approach to control of the manufacturing process that result in products with the desired quality attributes. Understanding variation and knowing how to detect and control it is therefore a key element to maintaining robust processes and systems.

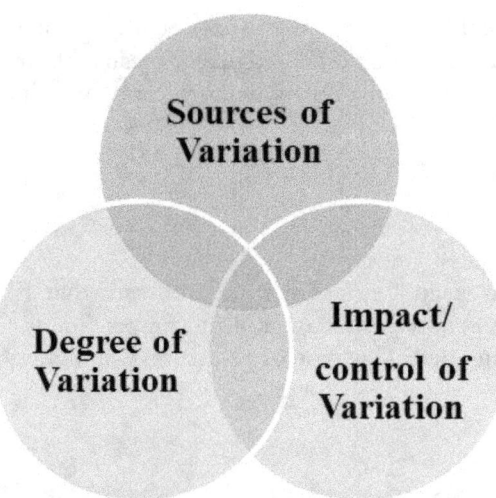

Figure 14: Variation

Manufacturers should, understand the sources of variation, the degree of variation, the impact of variation and how to detect and control variation. Each manufacturer should judge whether it has gained sufficient understanding to provide a high degree of assurance in its manufacturing process to justify commercial distribution. Focusing exclusively on qualification efforts without also understanding the manufacturing process and associated variations may not lead to adequate assurance of quality. After establishing and confirming the process, manufacturers must maintain the process in a state of control over the life of the process, even as materials, equipment, production environment, personnel, and manufacturing procedures change. Manufacturers should use ongoing programs to collect and analyze product and process data to evaluate the state of control of the process. These programs may identify process or product problems or opportunities for process improvements that can be evaluated and implemented through some of the activities described in Stages 1 and 2.

Manufacturers of legacy products can take advantage of the knowledge gained from the original process development and qualification work as well as manufacturing experience to continually improve their processes. Implementation of the recommendations in this guidance for legacy products and processes would likely begin with the activities described in Stage 3.

Stage 1 — Process Design: Process design is the activity of defining the commercial manufacturing process that will be reflected in planned master production and control records.

The goal of this stage is to design a process suitable for routine commercial manufacturing that can consistently deliver a product that meets its quality attributes. Building and Capturing Process Knowledge and Understanding Generally, early process design experiments do not need to be performed under the CGMP conditions required for drugs intended for commercial manufacturing and supply.

Stage 2 (process qualification) and Stage 3 (continued process verification). They should, however, be conducted in accordance with sound scientific methods and principles, including good documentation practices. Decisions and justification of the controls should be sufficiently documented and internally reviewed to verify and preserve their value for use or adaptation later in the lifecycle of the process and product. Although often performed at small-scale laboratories, most viral inactivation and impurity clearance studies cannot be considered early process design experiments.

Viral and impurity clearance studies intended to evaluate and estimate product quality at commercial scale should have a level of quality unit oversight that will ensure that the studies follow sound scientific methods and principles and the conclusions are supported by the data.

Product development activities provide key inputs to the process design stage, such as the intended dosage form, the quality attributes, and a general manufacturing pathway. Process information available from product development activities can be leveraged in the process design stage. The functionality and limitations of commercial manufacturing equipment should be considered in the process design, as well as predicted contributions to variability posed by different component lots, production operators, environmental conditions, and measurement systems in the production setting. Design of Experiment (DOE) studies can help develop process knowledge by revealing relationships, including multivariate interactions, between the variable inputs (e.g., component characteristics or process parameters) and the resulting outputs (e.g., in-process material, intermediates, or the final product).

Risk analysis tools can be used to screen potential variables for DOE studies to minimize the total number of experiments conducted while maximizing knowledge gained.

These activities also provide information that can be used to model or simulate the commercial process. Computer-based or virtual simulations of certain unit operations or dynamics can provide process understanding and help avoid problems at commercial scale. It is important to understand the degree to which models represent the commercial process, including any differences that might exist, as this may have an impact on the relevance of information derived from the models. It is essential that activities and studies resulting in process understanding be documented. Establishing a Strategy for Process Control Process knowledge and understanding is the basis for establishing an approach to process control for each unit operation and the process overall.

Strategies for process control can be designed to reduce input variation, adjust for input variation during manufacturing (and so reduce its impact on the output), or combine both approaches. Process controls address variability to assure quality of the product. Controls can consist of material analysis and equipment monitoring at significant processing. Decisions regarding the type and extent of process controls can be aided by earlier risk assessments, then enhanced and improved as process experience is gained. The planned commercial production and control records, which contain the operational limits and overall strategy for process control, should be carried forward to the next stage for confirmation.

Stage 2 — Process Qualification During the process qualification (PQ) stage of process validation, the process design is evaluated to determine if it is capable of reproducible commercial manufacture.

This stage has two elements: (1) design of the facility and qualification of the equipment and utilities and (2) process performance qualification (PPQ). During Stage 2, CGMP-compliant procedures must be followed. Successful completion of Stage 2 is necessary before commercial distribution. Products manufactured during this stage, if acceptable, can be released for distribution.

Design of a Facility and Qualification of Utilities and Equipment Proper design of a manufacturing facility is required under part 211, subpart C, of the CGMP regulations on Buildings and Facilities. It is essential that activities performed to assure proper facility design and commissioning precede PPQ.

Here, the term qualification refers to activities undertaken to demonstrate that utilities and equipment are suitable for their intended use and perform properly. These activities necessarily precede manufacturing products at the commercial scale. Qualification of utilities and equipment generally includes the following activities:

➢ Selecting utilities and equipment construction materials, operating principles, and performance characteristics based on whether they are appropriate for their specific uses.

➢ Verifying that utility systems and equipment are built and installed in compliance with the design specifications (e.g., built as designed with proper materials, capacity, and functions, and properly connected and calibrated).

➢ Verifying that utility systems and equipment operate in accordance with the process requirements in all anticipated operating ranges. This should include challenging the equipment or system functions while under load comparable to that expected during normal operation.

It should also include the performance of interventions, stoppage, and start-up as is expected during routine production. Operating ranges should be shown capable of being held as long as would be necessary during routine production. Qualification of utilities and equipment can be covered under individual plans or as part of an overall project plan.

The plan should consider the requirements of use and can incorporate risk management to prioritize certain activities and to identify a level of effort in both the performance and documentation of qualification activities.

Design of facilities and the qualification of utilities and equipment, personnel training and qualification, and verification of material sources (components and container/closures), if not previously accomplished.

Review and approval of the protocol by appropriate departments and the quality unit.. PPQ Protocol Execution and Report Execution of the PPQ protocol should not begin until the protocol has been reviewed and approved by all appropriate departments, including the quality unit. Any departures from the protocol must be made according to established procedure or provisions in the protocol. Such departures must be justified and approved by all appropriate departments and the quality unit before implementation (§ 211.100).

The commercial manufacturing process and routine procedures must be followed during PPQ protocol execution (§§ 211.100(b) and 211.110(a)). The PPQ lots should be manufactured under normal conditions by the personnel routinely expected to perform each step of each unit operation in the process. Normal operating conditions should include the utility systems (e.g., air handling and water purification), material, personnel, environment, and manufacturing procedures. A report documenting and assessing adherence to the written PPQ protocol should be prepared in a timely manner after the completion of the protocol.

This report should:

  ➢ Discuss and cross-reference all aspects of the protocol.
  ➢ Summarize data collected and analyze the data, as specified by the protocol.
  ➢ Evaluate any unexpected observations and additional data not specified in the protocol.
  ➢ Summarize and discuss all manufacturing non-conformances such as deviations, aberrant test results, or other information that has bearing on the validity of the process.
  ➢ Describe in sufficient detail any corrective actions or changes that should be made to existing procedures and controls.
  ➢ State a clear conclusion as to whether the data indicates the process met the conditions established in the protocol and whether the process is considered to be in a state of control. If not, the report should state what should be accomplished before such a conclusion can be reached. This conclusion should be based on a documented justification for the approval of the process, and release of lots produced by it to the market in consideration of the entire compilation of knowledge and information gained from the design stage through the process qualification stage.
  ➢ Include all appropriate department and quality unit review and approvals.

Stage 3 — Continued Process Verification

The goal of the third validation stage is continual assurance that the process remains in a state of control (the validated state) during commercial manufacture. A system or systems for detecting unplanned departures from the process as designed is essential to accomplish this goal. Adherence to the CGMP requirements, specifically, the collection and evaluation of information and data about the performance of the process, will allow detection of undesired process variability.

Evaluating the performance of the process identifies problems and determines whether action must be taken to correct, anticipate, and prevent problems so that the process remains in control (§ 211.180(e)). An ongoing program to collect and analyze product and process data that relate to product quality must be established (§ 211.180(e)).

The data collected should include relevant process trends and quality of incoming materials or components, in-process material, and finished products. The data should be statistically trended and reviewed by trained personnel. The information collected should verify that the quality attributes are being appropriately controlled throughout the process. We recommend that a statistician or person with adequate training in statistical process control techniques develop the data collection plan and statistical methods and procedures used in measuring and evaluating process stability and process capability.

Procedures should describe some references that may be useful include the following:

> ASTM E2281-03 "Standard Practice for Process and Measurement Capability Indices,"

> ASTM E2500-07 "Standard Guide for Specification, Design, and Verification of Pharmaceutical and Biopharmaceutical Manufacturing Systems and Equipment,"

> ASTM E2709-09 "Standard Practice for Demonstrating Capability to Comply with a Lot Acceptance Procedure."

Production data should be collected to evaluate process stability and capability. The quality unit should review this information. If properly carried out, these efforts can identify variability in the process and/or signal potential process improvements. Good process design and development should anticipate significant sources of variability and establish appropriate detection, control, and/or mitigation strategies, as well as appropriate alert and action limits. However, a process is likely to encounter sources of variation that were not previously detected or to which the process was not previously exposed. Many tools and techniques, some statistical and others more qualitative, can be used to detect variation, characterize it, and determine the root cause.

Leading manufacturers should use quantitative, statistical methods whenever appropriate and feasible. Scrutiny of intra-batch as well as inter-batch variation is part of a comprehensive continued process verification program under § 211.180(e).

Best practices ensures continued monitoring and sampling of process parameters and quality attributes at the level established during the process qualification stage until sufficient data are available to generate significant variability estimates. These estimates can provide the basis for establishing levels and frequency of routine sampling and monitoring for the particular product and process.

Monitoring can then be adjusted to a statistically appropriate and representative level. Process variability should be periodically assessed and monitoring adjusted accordingly. Variation can also be detected by the timely assessment of defect complaints, out-of specification findings, process deviation reports, process yield variations, batch records, incoming raw material records, and adverse event reports.

Production line operators and quality unit staff should be encouraged to provide feedback on process performance. We recommend that the quality unit meet periodically with production staff to evaluate data, discuss possible trends or undesirable process variation, and coordinate any correction or follow-up actions by production.

Data gathered during this stage might suggest ways to improve and/or optimize the process by altering some aspect of the process or product, such as the operating conditions (ranges and set-points), process controls, component, or in-process material characteristics. A description of the planned change, a well-justified rationale for the change, an implementation plan, and quality unit approval before implementation must be documented (§ 211.100). Depending on how the proposed change might affect product quality, additional process design and process qualification activities could be warranted.

Maintenance of the facility, utilities, and equipment is another important aspect of ensuring that a process remains in control. Once established, qualification status must be maintained through routine monitoring, maintenance, and calibration procedures and schedules (21 CFR part 211, certain manufacturing changes may call for formal notification to the Agency before implementation, as directed by existing regulations (see, e.g., 21 CFR 314.70 and 601.12).

The equipment and facility qualification data should be assessed periodically to determine whether re-qualification should be performed and the extent of that re-qualification. Maintenance and calibration frequency should be adjusted based on feedback from these activities.

Equipment Validation

Validation is a legal and regulatory requirement for the manufacture of medicinal products. The area of Validation can be sub-divided into two elements. Equipment Qualification (EQ) and Process Validation. Equipment qualification ensure that equipment operates as intended and is installed in accordance with the manufacturers recommendation. Process Validation involves the provision of documented evidence to confirm a particular process performs consistency and meets pre-determined specifications.

All equipment that can impact the quality of product is subject to Validation, hence equipment and systems used in aseptic manufacturing must undergo equipment and process validation. Installation Qualification (IQ) protocols should cover verification that all utilities are installed correctly to the manufacturers recommendations. All sitting and mechanical connections should also be confirmed as adequate. Other key tests and verifications includes:

- Documentation of Materials of Construction (MOC)
- Calibration of equipment based instrumentation
- Spare parts listing
- Preventative maintenance schedule creation
- Electrical installation verification
- Health and Safety assessment
- Ergonomic Assessment
- Documenting Software and Hardware
- Backup of software
- Backup of Recipes (Sterilization, Bio-decontamination etc.)

The system User Requirements Specification (URS) should provide the basis of testing and must be fulfilled during the course of Validation.

The ultimate goal of Equipment Qualification is to ensure that equipment is fit for its intended use. Therefore, equipment is validated to confirm it functions as intended and meets all requirements to manufacture product safely and consistently. FDA requires that "Each manufacturer shall ensure that all equipment used in the manufacturing process meets specified requirements and is appropriately designed, constructed, placed and installed to facilitate maintenance, adjustment, cleaning and use". In other words all manufacturing equipment, support facilities, measuring and test equipment shall be "qualified". (FDA 21 CFR 820.70 (g))

Equipment Qualification Protocols are developed to document this testing and hence provide evidence on the functionality and consistency of the equipment. There are two distinct parts within the scope of Equipment Qualification, Installation Qualification and Operational Qualification. Often these subparts are abbreviated to IQ and OQ. Other combinations such as IOQE and IQ/OQ can be encountered within industry. This is often defined in a company's procedure or SOP relating to Equipment Validation.

A User Requirements Specification (URS) is often used to document the "specified requirements" of a particular piece of equipment. A URS can then be used as an input document when equipment qualification is required. While a URS document can be extensive covering areas such as equipment functionality, utility requirements, safety features, software specs etc. not all requirements documented in a URS will need to be verified or validated. Critical requirements should be identified early and should always be verified.

In short Equipment Qualification is confirmation via documented evidence that the particular requirements for a specific intended use can be consistently fulfilled under anticipated conditions.

Often referred to as the three "C" s of Validation – confirmation – consistency and conditions (anticipated). These are key themes that Validation must address. Confirmation is addressed by the process of completing a formal validation. When it's done, it is documented and available for review to auditors. To assess consistency, there must be a number of batches or "runs". Typically, there is minor batch-to-batch differences or variations between batches. These differences can be as a result of setup or raw material differences. Process Validation must ensure that despite minor changes, there is consistency between batches, with product meeting specifications. Controlled or anticipated conditions are the machine or process settings that are known, documented and controlled during the manufacture of products.

Materials of Construction (MOC)

The materials of construction and evidence of the same (certificates) forms part of Installation also. Materials must be fit for the intended purpose and compatible with products and manufacturing agents that come into contact with them.
For example, fermenters are made of materials that are suited to the use of steam sterilisation techniques and regular cleaning. Such materials can be classed as both non-reactive and non-absorptive surfaces. Most aseptic processing equipment that incorporates product contact surfaces are made of high grade stainless steel. Cheaper classifications of stainless steel can be used for jacketing and other non-product contact areas.
All interior product contact surfaces should be polished to a "mirror" finish. Welds also need to be finished in a similar manner. Electro polishing provides a better quality surface finish than mechanical polishing.
 As with any chemical reaction, factors such as temperature, pH and oxygen concentration can impact the performance and yield. To ensure the optimum conditions are maintained, it is important to monitor and control such parameters and factors. By far the most common these days is automatic control of systems and equipment with automatic feedback and adjustment.

Operational Qualification (OQ) is the second component of Equipment Qualification. This is "Establishing by documented evidence that the equipment operates per specifications and over the required ranges and to required tolerances". Equipment is also tested to ensure alarms and controls operate as required and intended. Some typical checks included in an equipment-operational qualification are testing of alarms, control system testing, utility failures and functional and operational testing.

Suggested IQ/OQ Verifications / Tests

Standing Operating Procedures (SOPs)

SOPs are designed to provide formal documented instruction on how to execute tasks or operate equipment or machinery. While each company will require different headings, a work instruction or SOPs should cover set-up, system operation, cleaning and shutdown to name but a few.

Test Instrumentation Calibration

External test devices such as temperature probes, volt meters, lux meters and particle counters may be required to take measurements during an equipment qualification. Test instrumentation should have a suitable range, resolution and accuracy. Certificates of calibration should also be available, with calibration conforming to a recognized external standard. Information such as the serial number, model number and manufacturer should be recorded for reference. And traceability.

Equipment Based Instrumentation Calibration

All equipment based instruments must be calibrated as part of equipment validation or in advance of it. Instruments should have unique calibration ID.

Electrical Checks

Appropriate connections and earthing checks. Review of electrical drawings to ensure the physical status is as per drawings and specifications. Cables and electrical hazards should be also appropriately labelled.

Mechanical Checks

Ensure the systems are fixed, fastened and integrated mechanically. Safety guards and barriers should also be in place where required.

Pneumatic Checks

Verify the proper supply and integration of compressed air. Supply should be leak free, regulated with filters and watertraps fitted as required.

Documentation

Verification that design, operation and maintenance documentation has been received from the manufacturer and are stored appropriately.

Ergonomics

Controls and HMIs should be positioned to facilitate ease of use and should be identified clearly.

Health and Safety

Hazards are identified and guarded, pin points are identified. No trip hazards are evident and Emergency stops function.

Software

Equipment Installation software checks should record the names and version numbers of all software. HMI Software, PLC software, application software. Provision should be made for disaster recovery and backup.

Hardware

Computer hardware should be recorded to include the model, manufacturer, serial number and specification details.

Environmental

Any features detailed in the URS relating to environmental requirements need to be verified during IQ/OQ. For example, automatic shutdown after periods of inactivity. Heating and cooling systems should also be appropriately insulated.

Alarms

Automated processes such as sterilization tunnels, Autoclaves, Filling machines and Isolators typically have many alarms and controls. Alarms can be categorized as critical or non-critical to the process or product. Depending on the vendor or manufacturer alarms can also be grouped according to the type of alarm (EHS, process, mechanical, pneumatic and so on) Alarms should be tested to ensure the right action by the machine is taken, the process comes to a safe stop, and that the alarm can be acknowledged and the alarm condition cleared.

Utility Failure

Also referred to as provoke testing, utility failure of compressed air, fume extraction, electrical supply and so is to ensure in the event of failure during commercial manufacturing, the equipment comes to a safe stop and can be brought back into use upon recovery of the utility.

<u>Fixtures</u>

Materials of construction must be suitable for the intended use. In aseptic processing, high grades of stainless steel (316L) are the preferred material of use.

<u>Functional Tests</u>

The individual functions of equipment must be verified during commissioning and qualification.

Suggested Pre-Requisites to Equipment Qualification

Prior to formal Equipment Installation and Operational Qualification (IQOQ), there are a number of engineering activities that can be completed. Although work is required upfront in order to complete these activities such as preparing engineering test protocols, they will benefit the qualification stage. The completion of some or all of the above activities will help identify issues prior to formal equipment qualification. Essentially, an FAT Protocol is like an early draft of an IQOQ-Equipment Protocol.

Factory Acceptance Testing (FAT)

FAT or Factory Acceptance Test – is an Engineering activity, the purpose of the FAT is to verify the equipment or system meets the requirements of the URS. From the Validation Engineers perspective, it can be a learning activity and an opportunity to gather data, documentation and supporting design documents that will prove valuable during the equipment and Process Validation of the equipment. SAT is an Engineering activity that is completed at the site of the vendor or equipment manufacturer, post FAT.

Site Acceptance Testing (SAT)

Site Acceptance testing is also an engineering activity conducted when equipment arrives onsite. Depending on the company, SAT can be completed by the vendor or by the purchasing company. It consists of a series of installation and operational checks to ensure the equipment has not suffered any damage or deterioration between the disassembly, crating, shipping and delivery of the equipment.

Equipment Qualification (EQ) Protocols

<u>Protocol Preparation</u>
Thorough and careful preparation is critical in order to successfully complete a qualification without deviations. In the preparation of the protocol, the URS is a key document. Often quotations, design documents, vendor drawings and owner manuals can contribute to the test and verifications to be completed during EQ.

Protocol Approval

Approval is always required prior to executing an equipment qualification protocol. Approvers should be aware that they are signing for the accuracy and content of the whole document. It is strongly advised that prior to final approval and execution of a protocol, a dry-run or trial is completed to ensure the test methods and acceptance criteria are accurate.

Post Execution Review

Upon execution of a protocol, timely review is advised in order to catch any errors or omissions. The person reviewing the protocol should not be the same person who performed the test. This review is best completed by a Quality Engineer; however, each organisation should identify personnel responsible for EQ reviews.

Some points to remember when completing protocols:
- ➢ Ensure the protocol is fully approved prior to execution
- ➢ Ensure personnel are trained on the protocol (if required) and trained to the specific work instructions
- ➢ Ensure all team members, contractors etc. sign the signature log
- ➢ Observe safety precautions and wear PPE as required
- ➢ Ensure other employees that may be impacted by qualifications are aware that a qualification is in progress.
- ➢ Ensure that any test product required in support of the qualification is identified segregated and stored to internal standards
- ➢ Check that accurate work instructions are available (some companies may allow redlined copies to be used)
- ➢ Complete all tests in the protocol
- ➢ Always use indelible ink
- ➢ Carefully check each result against any acceptance criteria
- ➢ Ensure all test equipment is validated within calibration prior to use
- ➢ Handwritten comments should be signed and dated per GDP
- ➢ Deviations should be written up at the time of observation
- ➢ Data records and attachments should be identified with the protocol number, signed, paginated and dated
- ➢ When data is transcribed it should be verified by a second person. The source of the data should also be recorded
- ➢ All product manufactured should have relevant batch documentation as per normal production conditions

Equipment Qualification Reports

On completion, Equipment Installation and Operational Protocol reports are required. The format of any report largely depends on company specific procedures. If the protocol is an executable document (results are hand written in) then the executed version can be deemed the report. A summary report may be required but this depends on the requirements within your company or organisation.

The typical requirements of a completed EQ validation include:

> - Equipment Qualification Protocol
> - Equipment Qualification Protocol (executed)
> - Raw Data
> - Attachments (Examples of attachments include –material certs, calibration certs, CE Cert and MSDS'.

Software Validation

Where there is the potential to affect product conformance to requirements or where software or IT systems provide support to aspects of Quality Management, validation is required.

Most companies categorise software validations to account for the different applications of software and IT systems. For example, Enterprise systems, such as the drawing package SolidWorks would be validated in a different manner to Manufacturing Systems that contain software (a.k.a. embedded software).

"Embedded" software is where the software is integrated into the manufacturing equipment. Embedded software is typically validated during the Equipment Qualification stage, Process Validation stage or Test Method Validation. Enterprise software falls outside of Equipment or Process Validation but does require validation if it impacts product quality or is used to make quality decisions. Standalone systems such as ERP (Enterprise Resource Planning) systems also require validation.

Software Validation & GAMP

Good Automated Manufacturing Practice (GAMP) is a set of guidelines for manufacturers and users of automated systems in regulated industries. Specifically, the Medical device, pharmaceutical and biopharmaceutical industries. The application of GAMP and Validation of Automated Systems in manufacturing helps ensure that regulated medical devices and medicinal products have the required quality and are manufactured according to Good practices, meet regulatory and legal requirements and ensure patient safety. GAMP ensures quality is in-built into each stage of the manufacturing process. Therefore, GAMP has a place in all aspects of automation and production, including the handling of raw materials, control of facilities and equipment etc.

Key Terms

Automated System: Term used to cover a broad range of systems, including automated manufacturing equipment, control systems, automated laboratory systems manufacturing execution systems and computers running laboratory or manufacturing database systems. The automated system consists of the hardware, software and network components, together with the controlled functions and associated documentation. Automated systems are sometimes referred to as computerised systems; in this Guide the two terms are synonymous.

Commercial off-the-shelf (COTS): Configurable Programs- Stock programs that can be configured to specific user applications by "filling in the blanks", without (COTS) altering the basic program.

Computer System Validation: a process that confirms by examination and provision of objective evidence that the computer system conforms to user needs and intended uses. System validation is a process for achieving and maintaining compliance with GxP regulations and fitness for intended use by adoption of life cycle activities, deliverables, and controls.

GAMP 5: is a set of guidelines that offers a Risk-Based approach to ensuring the compliance of GxP impacting computerised systems.

V- Model: is a development process which sets out a roadmap of stages and deliverables during a project.

21 CFR Part 820: FDA requirements pertaining to Medical Devices.

User Requirement Specification, URS: The URS is a critical document that defines the requirements of the computerised system and agreement to the requirements.
Software Requirement Specification, SRS: an SRS can be written to interpret the requirements of a URS and how they relate to the requirement or how the requirement is met in practical terms regarding software.

Functional Design Specification, FDS: a functional design specification is a document that specifies how particular requirements are met – this can be a combination of how the equipment/process operates mechanically/automatically etc. An FDS is typically written to response to a URS.

Computer System Validation Life Cycle
The Computer System Validation Life Cycle refers to all activities from initial concept to retirement of a computer system. The life-cycle of the system includes the defining of, and performance of activities in a systematic way from conception, requirements, development or configuration, testing, release and operational use.

The four GAMP Life-cycle phases include:

> Concept
> Planning and Project stage
> Operation
> Retirement

The Concept Stage is concerned with understanding the need or the problem to be addressed. We will see that the User Requirement Specification (along with other specifications) and the initial risk assessment help to drive a project forward in a systematic manner. The most common life-cycle approach for Computerised and Automated systems is the V-Model. The GAMP based V-model lays out a roadmap which facilitates the Validation of equipment and automated systems.

The Planning and Project stage involves the planning of the validation effort required to implement the system into the business area(s) based on identification and approval of system concept. This phase includes assessments of the regulatory and system risks, supplier assessment, development of validation strategies, identification of deliverables that will be generated, definition of the business process the system will support as well as the user requirements which the system will fulfil.

Design & Development and configuration of the hardware and software is also required to meet the system requirements as per specifications. In case of custom Software components, this effort could also include detailed Software design and developmental testing to ensure readiness for verification testing.

Verification – This effort confirms that specifications have been met and releases the system for use. This phase will involve multiple stages of reviews and testing depending on the system type, the development method applied and its use. Once verification activities have begun any changes to the system must be captured through change control.

On successful completion of the verification activities, the system is then released for effective use. The Test strategy and other verification activities will vary widely between simple equipment and more complex customised/ configurable systems. The verification and validation approach is typically agreed and detailed in at the validation planning stage. The VP can be updated accordingly as the project develops with more detail been added. Alternatively, a test strategy document or matrix could be written to provide more specific test plans.

Verification deliverables vary based on the complexity and level or customisation of the system in question. Corporate or company specific procedures also shape the required activities to be completed and reported. Some generic deliverables are listed below.

> Approval, executing and review of test protocols
> Writing and approving SOPs for operation and maintenance of the system
> Traceability Matrix
> Completion of any Risk mitigations (e.g. updates to FMEA etc.)
> Validation Summary Report(s)

Validation reporting requirements varies depending upon the scope of the system and should also be driven by a procedure and template. The Validation Plan can also outline the deliverables and what needs to be addressed in the report. A Validation Summary Report (VSR) shall be written which summarizes the results of executing the VP the documents created for the validation activities, summarizes (or points to summaries) of the testing performed. Finally, the VSR indicate the acceptance of the system/equipment by the user by the Project team and state that the equipment is released for commercial operation / production.

The operation phase supports the need to maintain compliance and fitness for intended use after the system is released for normal use. It is important to ensure the system remains within a continued validated state. All proposed or necessary changes to the system must be assessed and controlled as part of a change control process. Once the system has been accepted and released for use, the operation phase begins. This phase consists of maintaining the system's compliant state and fitness for intended use through the control of the procedures supporting the system's operational use.

During the operation phase the below activities are typically completed:

- Ongoing Training
- Preventative Maintenance
- Service management and performance monitoring.
- Change Control
- Periodic review
- Maintaining system security
- Records management
- Calibration

The retirement phase involves the planning and proper management of activities relating to the removal of systems from service (shutdown). The retirement should take into account the storage of any data and any data migration that needs to occur prior to retirement. The retirement plan, if needed, will outline the retirement strategy from the roles and activities that will be conducted to the removal of the system for use. A Retirement Summary Report is produced that documents the results of the activities defined in the retirement plan including:

- Retirement Plan and Timelines.
- Summaries of any data migration activities.
- Identification of the storage location of documentation relating to the system.
- Obsoleting of SOPs.

It must be stressed that GAMP is a set of principles, a set of guidelines that aim to achieve compliant computerized systems that are fit for intended use. GAMP Guidelines differ to 21 CFR QSR regulations as they are not legal or statutory requirements. However, they represent industry best practice and compliment the Validation efforts that are legal requirements and statutory requirements.

Regulatory Review

Software Validation is a requirement of the Quality System regulation, 21 Code of Federal Regulations (CFR) Part 820. Validation requirements apply to:

(1) software used as components in medical devices,
(2) software that is itself a medical device, and
(3) software used in production of the device or in implementation of the device manufacturer's quality system.

Note: EU GMP Annex 11, provides information on the inspection of 'Computerised Systems'.

In addition, computer systems used to create, modify, and maintain electronic records and to manage electronic signatures are also subject to the validation requirements. Such computer systems must be validated to ensure accuracy, reliability, consistent intended performance, and the ability to discern invalid or altered records. The regulated user should be able to demonstrate through the validation evidence that they have a high level of confidence in the integrity of both the processes executed within the controlling computer system and in those processes controlled by the computer system within the prescribed operating environment.

Specification Hierarchy

An equipment URS can define the requirements of a computerised or automated system along with the operational, functional, process and safety mandated by the customer. If the system is bespoke and complex, an SRS may be written to more clearly detail the software requirements, automation and functionality. Similarly, an FDS (Functional design specification) can be written to address how mechanical or physical processing occurs.

URS-to-SRS

Scenario: a URS is written to specify the requirements for an automated blister packaging line in a Medical device company.
The URS details the following:

URS R1.0 – The machine shall be capable of operating in various modes to allow the manufacture of product and other debugging activity.

In turn, an SRS can be written to interpret the URS, for example, SRS-the system shall have the following modes:

SRS 1.1 Run empty mode -in this mode the equipment does not accept any new product.
SRS 1.2 Production mode – every station operates within the machine.
SRS 1.3 Bypass mode - where any operation can be disabled.

Another 2 examples of a URS requirement been transposed to an SRS requirement are shown below:

Example 1: URS Requirement

URS R1.0 In the event of E-Stop activation, all sequencers shall be maintained and shall retain the sequence step that they were in at the time of E-Stop activation.

SRS Requirement

SRS 1.0 The lot count and Lot integrity must be maintained after E-stop activation.

Example 2: URS Requirement

URS R1.0 The system shall use password protection
SRS Requirement

SRS 1. 1 Basic machine functionality (Cycle Start/Stop, Fault Reset, Manual Operations) require no security.

SRS 1.2 As required, User IDs will be assigned to security groups for authentication. Authentication will be via Active Directory authentication against domain accounts.

SRS 1.3 An auto-logoff feature shall be incorporated in the design.

Examples of Security requirements:

- Three levels of access required, operator, and engineer and maintenance
- Engineer - access to all screens, to modify process settings Maintenance - access to functions required to perform machine maintenance activities.
- Operator - restricted access, does not have access to change process settings.
- Different access levels will require different passwords.
- No security will be required for basic operations (Start/Stop)
- A user auto-logoff feature shall be incorporated in the design. The auto- logoff time shall be configurable.
- A soft copy of Program settings must be provided with delivery of the equipment.

Examples of Security requirements:

- The reject count and yield must be displayed on the HMI screen.
- Real-time readings for all critical parameters shall be visible on the HMI Screen.
- All Critical parameters shall be adjustable via the HMI Screen.
- The status of each door should be visible on the HMI screen.

Examples of EHS requirements:

- Activated E-Stops shall be clearly displayed on the HMI screen with a suitable alarm message generated.
- Activated E-Stops shall result in no further movement of the system until the E-stop is reset and all alarms are cleared.

System Categorisation

GAMP 5 makes provision for four categories of software in order to distinguish the level of customization/configurability that exists across software's serving different functions.

GAMP Software Category 1, Operating Systems
GAMP Software Category 2, Non configured software
GAMP Software Category 4, Configurable software packages
GAMP Software Category 5, Custom Software

GAMP Software Category 1, Operating Systems

Category 1, operating systems, covers established commercially available operating systems. These are not subject to validation themselves, the name and version of the operating system must, however, be documented and verified during Installation Qualification (IQ). Application software hosted on operating systems need to be validated.

GAMP Software Category 2, Non configured software

Category 3 covers commercially available, standard software packages and "off the-shelf" solutions for certain processes. The configuration of the software packages should be limited to adaptation to the runtime environment (for example network and printer connections) and the configuration of the process parameters. The name and version of the standard software package should be documented and verified in an Installation Qualification (IQ). Special user requirements, such as security, alarms, messages, or algorithms must be documented and verified in an Operational Qualification (OQ).

GAMP Software Category 4, Configurable software packages

GAMP Software Category 4, Configurable Software Packages Category 4 covers configurable software packages that allow special business and manufacturing processes. This involves configuring predefined software modules. These software packages should only be considered as belonging to Category 4 if they are well-known and mature. Normally, a supplier audit is necessary. If this is not available, the software packages should be handled as Category 5. The name, version, and configuration should be documented and verified in an Installation Qualification (IQ). The functions of the software packages should be verified in terms of the user requirements in an Operational Qualification (OQ). The Validation Plan should take into account the lifecycle model and an assessment of suppliers and software packages.

GAMP Software Category 5, Custom Software

GAMP Software Category 5, Custom Software Custom/Bespoke Software (GAMP Software Cat 5) is software that contains custom code designed or modified specifically for a particular customer. As the code is custom it presents a greater risk. This risk must be mitigated with the right approach to the validation.

GAMP Considerations

Correctly assigning a GAMP software category to equipment, a system or process is an important activity that should be completed early-on in the planning stage of a project. There must of some degree of familiarity with the equipment or system. The manufacturer or vendor can be a source of information that may help the designation. In many cases, companies create tools or processes that help determine what GAMP software category applies. These have different names such as questionnaires, screening tools, planning tools etc.

Risk Assessments

A Risk Assessment process should be applied to cGxP computerized systems in order to identify and mitigate potential risks to (1) patient safety, (2) product quality and (3) data integrity. Results identified through a Risk Assessment help to determine the validation strategy, the effort and time required, and allow better targeting of the validation activities to the highest risks.

The Risk Assessment should be revised during the Software Development Lifecycle (SDLC) if the functionality, requirements or intended use of the system changes. The Risk Assessment activity should also be evaluated during system build-up as well as when implementing changes. Risk Assessment tools for cGxP computerized systems are typically completed during the planning stage, specification stage and post qualification if a change or update is required.

Planning Stage

Initial Impact/Risk Assessment – during the planning phase to identify the level of impact and GxP relevance of the system/equipment. (Tools used: High Level Risk Assessment).

Specification Stage

Functional or Quality Risk Assessment – during the specification phase - identify potential risks and possible mitigations to be to be introduced to the process. (Tools used: Quality Risk Matrix, (p)FMEA).

Changes to the system

Impact Assessment of changes – as part of the change control process in the system operational phase. The following diagram defines the Risk Assessment steps within the System Life Cycle (Tools used: Impact assessment checklist, Change control procedures).

Quality Risk Matrix

A QRM is a risk assessment that identifies and manages the risk to patient safety, product quality and data integrity that relate to the systems processes. Risk Scenarios or potential causes should be developed for each identified function or process step and then assessed for the impact on patient safety, product quality or data integrity. Risk mitigations and controls should then be introduced to address both medium and high levels of risk. The QRM requires 3 "assessments" in order to produce an estimation or overall Risk (Low, medium, high)

Assess Likelihood
Assess Detectability
Assess Severity

Traceability Matrix

A Traceability Matrix should be prepared as required in accordance with company and internal policy. It is also recommended by GAMP guidelines, ASTM E2500 and ISPE Risk based approach to Validation. The matrix links the user requirements and specifications to the testing and validation activities. A traceability matrix illustrates that all user requirements are traceable to the verification/validation activity or vendor documents as relevant (FDS if applicable, Design specifications etc.) A simple traceability matrix (TM) format is shown below on the next slide. Generally, individual organisations will have an approved template to work from. However, the URS structure can form the basis of the template, with additional columns added to document the test/verification method, Reference documents (such as FDS' and vendor specifications and design documents)

Configuration Identification

Software and hardware packages should be identified by a unique product identifier and a version number. For the software end-user, the parts of an automated system that are subject to configuration management should be clearly identified. The system should therefore be broken down into configuration items. These should be identified at an early phase of development so that a complete list of configuration items is defined and maintained. The application-specific items should have a unique name or version ID. The depth of detail when specifying the elements is decided by the needs of the system, and the organization developing that system.

Requirements for the User ID and Password

User ID: The user ID of a system should have a minimum length agreed with the customer and should be unique within the system.

Password: A password should always consist of a combination of numeric and alphanumeric characters. When setting up passwords, the number of characters and a period after which a password expires should be stipulated. The structure of the password is normally selected to suit the specific customer. The configuration is described in the section Security Settings of Password Policy.
Criteria for the structure of a password are as follows:
Minimum length of the password
Use of numeric and alphanumeric characters
Case sensitivity

Audit Trail

The audit trail is a control mechanism of a system that allows all data entered or modified to be traced back to the original data. A reliable and secure audit trail is particularly important in conjunction with the creation, change or deletion of GMP relevant electronic records. In this case, the audit trail must archive and document all the changes or actions made along with the date and time. Typical contents of an audit trail must be recorded and describe the procedures "who changed what and when" (old value/new value).

Uninterrupted Power Supply

An uninterruptible power supply (UPS) is a system for buffering the main power supply. If the power supply fails, the battery of the UPS supplies the required power. When the power supply returns, the UPS battery stops supplying power and is recharged. Some UPS systems provide the option of main power supply monitoring in addition to the buffering function. They guarantee an output voltage at all times without interference voltages. UPS systems are necessary so that process and audit trail data can continue to be recorded during power failures. The design of the UPS must be agreed with the system user and must be specified in the URS, FS or DS. The following points must be considered:

> Energy requirements of the systems to be supplied
> Power of the UPS
> Required duration of UPS buffering

The energy requirements of the systems to be buffered decide the size of the UPS. A further selection criterion is the priority of the systems. Systems with high-priority include:

> Automation system (AS)
> Archive server
> Operator station (OS) server
> Operator station (OS) clients
> Network components

Field devices that generally have relatively high energy requirements may also be included in the buffering depending on the power of the UPS. This must be decided in consultation with the system user and related to the classification of the process. Whatever is decided, it is important that the systems for logging data are included in the buffering. The time at which the power failure occurred should also be recorded. The use of UPS systems involves the installation of software. This should be installed and configured on the PC-based computers of the process control system to be buffered. The setup should also account for:

> Configuration of the power failure alarms
> Stipulation of the time before the PC is shut down
> Stipulation of the time during which UPS buffering is provided

The automation systems (AS) must be programmed so that the process control system changes to a safe state after a selectable buffer time if a power failure occurs.

Types of UPS

Due to the different requirements of the various devices involved, three classes have established themselves as stipulated by the International Engineering Consortium (IEC) in product standard IEC 62040-3 and the European Union EN 50091-3:

Offline UPS

The simplest and least expensive UPS systems (according to IEC 62040-3.2.20, UPS class 3) are standby or offline UPS systems. They protect only against power outages and brief voltage fluctuations and peaks. Undervoltage and overvoltage are not compensated. Offline UPS systems switch to battery supply automatically if there is overvoltage or undervoltage.

Line-interactive UPS

The way in which line-interactive UPS systems (according to IEC 62040-3.2.18, class 2) function is similar to standby UPS systems. They protect against power outage and brief voltage peaks and can compensate voltage fluctuations continuously using filters.
Online UPS

Double conversion or online UPS systems (according to IEC 62040-3.2.16, class 1) count as genuine power generators that continuously generate their own line voltage. Connected consumers are therefore supplied permanently with line power without restrictions. At the same time, the battery is charged

Software-Source Code Review

For GAMP Software Categories 4 and 5 source code review is advised unless the supplier has evidence of the same available for review. As part of Good Automated Manufacturing Practices, reviews should be completed as part of the development lifecycle. If a source code review is not completed a justifiable rationale should be documented in an applicable document such as a Validation Master Plan.

Calibration

A key part of any qualification is to confirm that the equipment is fit for the intended purpose. Each piece of equipment will have a defined operating range. For example an oven may have an operating range of 20°C to 100°C ±5°C. However, the process window may only require a temperature range of 30°C to 60°C. In this instance a calibrated range of 20°C to 70°C would suffice. However, if the process window or the temperatures at which product was manufactured ranged from 20°C to 100°C this would present a problem as it falls outside of the equipment qualification range when the calibration tolerance is taken into account.

Deviations

A deviation can be simply described as an unintended event which causes a test or verification to fail to meet expected acceptance criteria. Each company or organisation should have a procedure detailing the management of deviations. It is critical that all deviations are identified, investigated and evaluated for their impact on product quality, the risk/impact to the patient and the impact on the qualification or validation. The basic components to a deviation are listed below:

- ➢ Deviation Description- provides the page and section of the deviation and an overall description eg. document generation error, operator error, machine crash etc.
- ➢ Potential impact on product – does the deviation impact the product?
- ➢ Potential impact on validation/qualification-will the validation have to be repeated in part or in full?
- ➢ Investigation- DMAIC, RCA, Fishbone Diagram, 5W
- ➢ Root Cause- what is the concluding root cause?
- ➢ Planned Resolution- what actions are required to be implemented?
- ➢ Deviation Resolution (Actions completed)- were all the actions in the planned resolution implemented? What is the final result? Have the actions been effective? Requalification

Over the lifetime of a piece of equipment, the need to requalify may arise. Therefore, any proposed change to equipment or a process must be assessed to see if the validated state will be impacted. It is therefore critical to understand clearly the nature of the change(s). Some scenarios where requalification of equipment may be required include:

- ➢ Major equipment repairs
- ➢ Moving equipment
- ➢ Changes to the upper and lower operating limits of the equipment
- ➢ Upgrading of software
- ➢ Hardware upgrades or changes
- ➢ Changes in performance and/or defect levels

After assessing any proposed changes based on the reasons listed above a determination of the level of requalification is required. This may be limited to a partial requalification (addendum) or it may require a full requalification.

☐

Process Validation

This section provides an introduction to Process Validation for medical devices. Process validation is a statutory and regulatory requirement for the manufacture of medical devices. Per FDA 21 Code of Federal Regulations Process Validation is a regulatory requirement of Good Manufacturing Practices (GMP) for both pharmaceuticals (21 CFR 211) and medical devices (21 CFR 820). In addition to the regulatory drivers, process validation is a requirement in order to obtain certification to international standards issued by many notified bodies. (e.g. ISO 13485 Medical Devices- Quality Management Systems, ASTM E2500-Standard Guide for Specification, Design, and Verification of Pharmaceutical and Biopharmaceutical Manufacturing Systems and Equipment etc.)

<u>Traditional and New Approaches to Validation</u>

Historically, process validation involved the testing and verification of all aspects of a process. While this may seem appropriate, it must be understood that in order to test/verify all aspects of a process, for it to hold weight, this activity must be documented and recorded. In this respect, an "all aspects" approach to process validation can be burdensome to resources. The traditional approach largely used the V-Model which set out a sequence of deliverables that should be completed. The use of risk assessments were limited as all requirements of a system were tested and qualified.

In recent years, a risk based approach has been increasingly endorsed by regulatory authorities and hence adopted by medical device manufacturers. One such standard is the ASTM E2500. As the title suggests, it is primarily used within pharmaceutical and biopharmaceutical industries, its principles and core approach can be adopted by medical device manufacturers also. ASTM E2500 was designed to make the implementation process for GMP systems and validation more cost-effective. It aims to achieve this based on scientific and risk-based principles, focusing on the risk to the patient. However, at just a five-page document, ASTM E2500 lacks the detail required in order to meet regulatory expectations. While different terminology and philosophies exist they do not change the regulatory expectations relating to validation. Both approaches exhibit common elements which include:

➤ Good engineering practices
➤ Planning
➤ Requirements definition (URS etc.)
➤ Design review
➤ Change Management
➤ Documented testing and inspection

While many manufacturers may predominantly choose a particular approach, it is common to see elements of both approaches (traditional & risk based). Each individual company will shape its internal validation procedures to best suit its business needs of the company.

What Is Process- Operational Qualification (OQ-P)?

The ability of a process to produce product in accordance with pre-determined specifications under worst case conditions. PQ is only required if no worst case conditions are evident.

What Is Process-Performance Qualification (PQ)?

The ability of a process to consistently produce product in accordance with pre-determined specifications under anticipated conditions (normal/routine conditions). Before considering Process Validation in further detail, it is important to look at the pre-requisites and other supporting activities required. These are examined in the sections below.

Test Methods & Process Validation

It is important to consider test methods early on in the validation lifecycle. Before you can begin to consider Process Validation, test methods should be understood and in place.

A Test Method is a process or an action used to verify that a product feature meets a predefined specification. Tests methods can be physical or analytical in nature. Test Method validation should be completed in advance of process validations to allow the proper assessment of process and product outputs meaning it is often a pre-requisite to Process Validation.

Examples of test methods include simple visual inspection by microscope, measurement of a dimension with a calipers or measurement of a dimension using an automated optical inspection system. Some test methods will involve MSA (Measurement System Analysis) studies for example, a measurement of a dimension by an operator using a microscope. In contrast, a test method to determine organic residuals would require an Analytical Test Method validation.

The equipment must be qualified (Installation Qualification and Operational Qualification) before the method is validated. Remember – Testing completed in contract laboratories or specialist services also require validation! Test methods are critical to the success and integrity of your Process Validation as they assess the outputs. E.g. what are the dimensions, physical attributes or chemical properties of the product and how do they confirm to specifications?

Fundamentals of Process Validation

The most important point when it comes to validation is that validation is neither exploratory nor investigative. Equally, it is not an engineering study. If you are ready to validate a system or process, all of the groundwork must be completed. This means critical parameters must be defined and documented, with technical rationale on why such parameters are critical etc. This body of work is typically done during a process development study or protocol. Validation of confirmation, so Process Validation is confirming that a process is capable of consistently manufacturing product under anticipated conditions. Remember, validation should be representative of the commercial process, so any issues in Process Validation will be repeated in commercial manufacturing.

Consistency, a core principle of Process Validation is typically demonstrated by producing 3 batches/runs for a Process Performance Qualification (PPQ). These batches should be representative of normal production i.e. the size of the batch should be typical of commercial volumes. The PQ study should be executed at nominal conditions, (often termed "anticipated conditions") essentially referring to a controlled environment. Controlled material and controlled parameters (CPPs) are required. Nominal settings should be selected for PQ.

Process Validation and Dominance Factors

The concept of dominance is a term used to describe the "influential" or "dominating" effect on a system or process. Typical examples include the injection moulding process, and packaging process. For example, an injection moulding process can be said to have material as a dominant factor. Batch to batch differences of resin or raw material may cause a change to outputs such as dimensions of a product or component. If dominant factors cannot be identified or understood a "Designed Experiment" (DoE) technique can be used to properly determine them. Dominance can be categorised into 5 sections. (1) Setup Dominance (2) Time Dominance (3) Worker Dominance (4) Information Dominance and (5) Component Dominance.

Setup Dominance

Setup Dominance -The Process or Equipment relies principally on a procedure or process setup. Process should be stable one "set-up".
Examples include ovens and package sealers. With regard to the oven, the setup would generally be controlled by a recipe or program. This program would be selected by the operator through the Human Machine Interface (HMI). The setup with the correct version of the recipe that contains the desired temperatures, times and pressures is therefore a critical input to the process. With regard to the Packaging Machine (blister packaging), the correct setup for the tooling and program are critical inputs. If Setup Dominance is significant, it is best practice to have 3 separate set-ups/changeovers in the Performance Qualification (PQ).

Time Dominance

The Process or Equipment is subject to changes over time (drift over time in temperature, solvent cleanliness, tool wear etc.). The process may need a schedule of process checks and adjustments to ensure process consistency.
Examples include CNC Machinery (tool wear) or aqueous based cleaning systems. The tool may only be able to manufacture 1000 parts before defects or quality issues are encountered. If Time Dominance is significant 3 time points or cycles of expected variation should be made e.g. 3 points in the cycle (start, middle and end) or 3 points in a shift (start of shift, middle of shift and end of shift).

Worker Dominance

For Worker dominance, the process requires operator experience and skill. Examples include manual or hand finishing. If Dominance is significant, ensure there are a minimum of 3 operators involved in the manufacturing/ activity.

Information Dominance

With information dominance, the process or Equipment requires the transmission and/or analysis of information. Examples include LIMS, MRP and ERP systems. A minimum of 3 information transmissions in the PQ should be completed.

Component Dominance

The Process is influenced by the variability of the input materials and/or components. It requires robust inspection and sorting procedures as well as process adjustments. When Component Dominance is significant, ensure there are a minimum of 3 component/raw material batches in the PQ sampling plan. If component dominance is significant this can be mitigated by including the material/component variation in "worst case" testing as part of the Operational Qualification Process (OQ-P)

Process Operational Qualification (OQ-P)

During the Operational Qualification-Process (OQ-P) study worst-case process conditions are normally employed. This may be worst case temperatures, speeds, feeds etc. The OQ-P should challenge the manufacture/processing of product at the limits of the processing window (process range). If no worst-case conditions exist, then an OQ may not be required and only a Performance Qualification is required.

A family or matrix approach is often used where similar products are to be validated. A particular product size of product configuration may be selected to represent the worst-case product. Therefore, by qualifying the worst case, all other products within that family of products would be considered validated. However this approach must be clearly documented and technical rationale provided in advance of any qualification activities. This can be addressed in a Validation Plan or within a protocol.

Protocol Approval Check list

The Validation Protocol is the means by which objective evidence is documented and gathered. The Validation Protocol is therefore a critical document. It should clearly set out the approach to the validation, detailing methods, tests and verifications to be completed and the acceptance criteria that applies to such tests and verifications. Remember, a validation document is a legal and regulatory document and can be subject to detailed scrutiny. Below are some suggested general checks to apply when writing Validation Protocols.

Author
- SOP available -Protocol conforms to validation procedure.
 - Ensure item numbers and batch size are correct.
-Test methods are correct.

SME Reviewer
- Is the protocol number correct?
-Review content of Protocol for accuracy and completeness.
 - Protocol conforms to validation procedures.

- Procedure and evaluation table are appropriate and correct.   Engineering:
- Review content of Protocol for accuracy and completeness.
- Specifications and operating parameters are correct.

QC / Laboratory
- Review content of protocol.
 - Raw Material Specifications are in place.
- Finished Product Specifications are in place.
- Testing and sample size is correct.

Quality
- Review content of protocol.
- Protocol conforms to SOPs.
 - Evaluation and acceptance criteria are appropriate.

Process Performance Qualification

The purpose of the PPQ is to demonstrate the capability of the process to consistently manufacture product to pre-determined specifications under normal operating conditions and defined parameters. Validation is confirmation, so process validation is confirming that a process is capable of consistently manufacturing product under anticipated conditions.
   ➢ Lots should be produced consecutively (in sequence)
   ➢ Lots must meet the acceptance criteria set out in the protocol
   ➢ The lot size should be reflective of the intended lot size and also take into account normal variation
   ➢ If a family approach or matrix approach is used, the product selection must be clearly justified and documented
   ➢ Execute under anticipated conditions, essentially this refers to a controlled environment. Controlled material, controlled parameters (CPPs)
   ➢ Nominal settings should be selected for PPQ

Yield Data (aka Process Yield Data)

Process Yield is a term used in manufacturing to represent the overall process performance. Yield is most often expressed as a percentage of good/passing product. It reports the % of compliant units, that is units or products that meet the product acceptance criteria (eg. CQAs). The remaining "bad" units are classified as defects or scrap. In some manufacturing processes, rework is possible or permitted.

Yield data often forms part of the acceptance criteria for a validation. The overall process yield for each batch should be calculated and compared to the starting process weights or units to determine loss due to processing as it is common to lose material during processing.

Once the initial validation is completed it is important that the system or process remains within the validated state, meaning that the system remains in a state of controlling process systems that capture information and data about the performance of the process. The use of statistical trending techniques should be considered. Data analysis of process and product should also include trending of raw materials, components and finished product. The purpose of process monitoring is to ensure critical parameters remain within control limits. It also helps to identify increasing variability or instability within the process which can then be investigated. All processes must have an upper and lower limit. If a process parameter only has a one-sided limit, then provide rationale in the OQ protocol to justify why a one-sided parameter window is acceptable. This requirement is not applicable to parameters that are set points.

Revalidation (or Maintaining a Validated State)

Revalidation is sometimes required if the original validation is no longer valid or representative of the process. Some instances where revalidation must be considered include changes to the process that can affect the product quality or efficacy, a removal or the addition of a processing step or transfer of the equipment to a different location. In many companies an Impact Assessment is conducted if there is a proposal to modify a manufacturing process. Some changes may not require any validation while others may require a verification run.

When changes are proposed to the validated state of a process, the proposed changes must be fully understood in terms of the impact to product quality and the validated state. A risk assessment should be conducted to determine risks and appropriate mitigations.

Scenarios on maintaining a Validation State

Line Addition-Product (New Product or Product Transfer)

This may be required if a new product has been introduced but uses the same process(es) for manufacturing. Typically this can apply if a new size has been introduced. For example, a new size of surgical blade.

Line Extension

A line extension commonly refers to a scenario where the product/process is different or considered outside the existing range or processing parameters.
Note: The impact on the validated state for line additions and line extensions should be assessed formally and documented. A line extension may require a new validation or addendum to the existing validation, whereas a line addition to add a new product may be within the scope of existing validations.

Acceptance Criteria

The acceptance criteria contained in Validation Protocols are normally based on established product specifications. For example, a contact lens manufacturing company may produce a lens with a diameter of 18mm ±0.2mm. The product produced during a process validation must be inspected to record the diameters of lenses being manufactured. Disposition of product is based on the product specification and determines if the product feature measured receives a pass or fail.

In addition to Product Specifications, it is common to have acceptance criteria such as Yield, and OEE. The acceptance criteria for these conditions are normally driven by an internal company procedure or alternatively can be detailed in the validation plan or protocol study.

Validation Strategies

A Family Approach (a.k.a. Bracketing, Matrix Approach) to validation is often used where a variety of similar products are manufactured using the same equipment. For a Process Validation, a product that is representative of the family or group of products may be selected. Alternatively, a 'worst case' product may be selected as it presents the greatest challenge to manufacture to product specifications.

Principles of Worst Case Selection

Worst Case is a particular condition, set of conditions, and/or set of process parameters, generally made up of processing limits. Worst case conditions present the greatest chance of process issues or the greatest chance of failures due to product quality. Worst case conditions are used at OQ-P stage to provide the greatest level of challenge, however, this is outside of normal operating conditions.

Requalification

During the lifetime or a process or piece of equipment, the need to re-qualify may arise. Such need should be assessed according to a validation procedure. Generally, the same tools used in the original validation can be re-applied to identify the need or re-qualify and indicate what requirements must be included.

The first step must be a review of the existing qualification, as changes may not impact the validated state, or may only require a limited requalification. For example, moving a piece of equipment may only require requalification of the utilities such as compressed air or process water if the operation of the equipment is not impacted by the movement and re-siting.
Some examples where re-qualification may be required include:

➢ Transferring a process from one plant to another plant
➢ Changes to the process settings which may impact the product quality
➢ Changes to the design of the product
➢ Changes to manufacturing aids (e.g. cleaning agents, jigs and fixtures)

CHAPTER 14

Change Control

Introduction

Change control can be defined as a formal system in which qualified representations of appropriate disciplines review changes that may affect the validated status of facilities, systems, equipment of processes and determine the need for action that would ensure and document that the system is maintained in a validated state. The term Change management is also widely used. It can be defined as the systematic approach to proposing, evaluating, approving, implementing and reviewing changes. (Ref: ICH Q10)

Note: WHO guidance and PIC/s does not specify any particular requirements in regard to Change Control.

Impact of Changes

Changes in processes, materials and equipment are unavoidable throughout the lifecycle of a product or process. However, in order to ensure changes do not have an adverse impact on product safety or quality. Proposed changes should be assessed by all stakeholders, typically involving review of the change by cross functional teams with a documented review of the change along with any tests or verifications required in order to implement the change safely.

<u>Tools in identifying Potential Impact</u>

6M/6P/4S are simple and effective tools to use in order to identify equipment, areas, processes or documentation that may be impacted by a planned change.

6 M	6 P	4 S
Machines Methods Mother Nature Materials Measurement Manpower	People Process Policy Plant Programs Products	Surroundings Suppliers Systems Skills

Figure 15: Lean tools

Templates can also be used to visually document 6M or 6P methodologies.

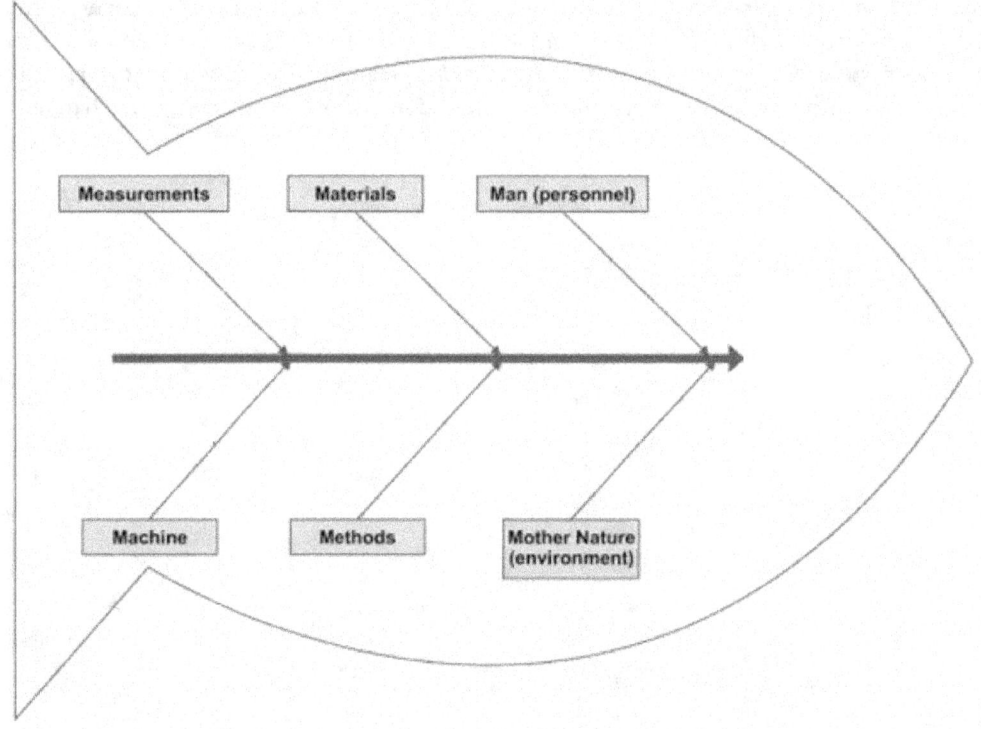

Figure 16: Fishbone visual representation of 6M.

Control of Changes

Written procedures should provide for the identification, documentation, appropriate review, and approval of changes in raw materials, specifications, analytical methods, facilities, support systems, equipment (including computer hardware), processing steps, labelling and packaging materials, and computer software.

CHAPTER 15

Complaints and Recalls

Introduction

Patients and medical professionals are encouraged to be vigilant and report any potentially defective or suspect products. Manufacturers must have a system in place to receive complaints and a written procedure that details how complaints are reviewed and acted upon. Manufacturers must also have a system that facilitates the recall of products known or suspected to be on the market.

Suspect or defective product

Complaint received

Assessment of complaint according to written procedure

Decision making

Recall (if required)

General points:

 ➢ Record all details of complaint

 ➢ Document the investigation

 ➢ Document all decisions

 ➢ Review type and amount of complaints regularly

 ➢ Communicate with Competent authorities as appropriate

Investigation

Q7 Guidance states the following with regard to investigations:

 ➢ *The information reported in relation to possible quality defects should be recorded, including all the original details. The validity and extent of all reported quality defects should be documented and assessed in accordance with Quality Risk Management principles in order to support decisions regarding the degree of investigation and action taken.*

 ➢ *If a quality defect is discovered or suspected in a batch, consideration should be given to checking other batches and in some cases other products, in order to determine whether they are also affected. In particular, other batches which may contain portions of the defective batch or defective components should be investigated.*

➤ *Quality defect investigations should include a review of previous quality defect reports or any other relevant information for any indication of specific or recurring problems requiring attention and possibly further regulatory action.*

➤ *The decisions that are made during and following quality defect investigations should reflect the level of risk that is presented by the quality defect as well as the seriousness of any non-compliance with respect to the requirements of the marketing authorisation/product specification file or GMP. Such decisions should be timely to ensure that patient and animal safety is maintained, in a way that is commensurate with the level of risk that is presented by those issues.*

➤ *As comprehensive information on the nature and extent of the quality defect may not always be available at the early stages of an investigation, the decision-making processes should still ensure that appropriate risk-reducing actions are taken at an appropriate time-point during such investigations. All the decisions and measures taken as a result of a quality defect should be documented.*

➤ *Where human error is suspected or identified as the cause of a quality defect, this should be formally justified and care should be exercised so as to ensure that process, procedural or system-based errors or problems are not overlooked, if present.*

➤ *Appropriate CAPAs should be identified and taken in response to a quality defect. The effectiveness of such actions should be monitored and assessed.*

➤ *Quality defect records should be reviewed and trend analyses should be performed regularly for any indication of specific or recurring problems requiring attention.*

Recalls

Recalls should be managed and co-ordinated by a responsible person with adequate support from a wider team to handle all aspects of the recall or complaint. This responsible person typically is independent of the sales and marketing organisation.

GLOSSARY

GLOSSARY

A

Accelerated Aging

When the deterioration of a device or product component from natural aging is accelerated and simulated in the laboratory.

Accuracy

Accuracy or trueness. An expression of the closeness of agreement between the value that is accepted, either as a conventional true value or an accepted reference value and the value obtained. A system with low bias implies good accuracy and vice versa.

Adverse Event

A situation or condition that occurs when a data point, result, or process etc. is outside the expected or predetermined limits or ranges.

Air Exchange Rate Per Hour (ACPH)

The rate of air exchange expressed as number of air changes per hour and calculated by dividing the volume of air delivered in the unit of time by the volume of space.

Active Pharmaceutical Ingredient

Any substance or mixture of substances intended to be used in manufacturing a drug (medicinal) product and that, when used in the production of a drug, becomes an active ingredient of the drug product. Such substances are intended to furnish pharmacological activity or other direct effect in the diagnosis, cure, mitigation, treatment, or prevention of disease, or to affect the structure and function of the body. (ICH Q7A, Annex 18, Part II)

ANSI

American National Standards Institute

Antimicrobial Resistance

Antimicrobial resistance corresponds to the emergence and spread of microbes that are resistant to cheap and effective first-choice, or "first-line" antimicrobial drugs.

Application

A term most often used in relation to Software validation and computerized systems. It is any software installed on a defined platform providing specific functionality.

Approve

"Approve" the device after reviewing a premarket approval (PMA) application that has been submitted to FDA.

AVL (Approved vendor list)
A list of all the vendors or suppliers approved by a company as sources from which to purchase materials.

Artwork

Electronic files or printouts containing the representation of a packaging item, graphical elements, and regulatory text. Approved artworks are used by suppliers for printing.

Aseptic (conditions)

Conditions in the working environment under which the potential for microbial and/or viral contamination is minimized.

ASTM
American society for testing and materials.

ATEX

ATEX, an acronym of the French Atmospheriques Explosives. This European Directive amends and adds safety requirements for hazardous areas in the relevant national legislation in the member states of the European Union, bringing in a common standard. Where equipment is to be used in potentially explosive atmospheres containing gas or combustible dust, it must comply with the ATEX directive.

Audit Trail

The audit trail is a control mechanism of a system that allows all data entered or modified to be traced back to the original data. A reliable and secure audit trail is particularly important in conjunction with the creation, change or deletion of GMP relevant electronic records.

Acceptable Quality Level (AQL)

The AQL of a sampling plan is the Process Performance Level routinely accepted by the sampling plan.

Basis of Design

A design document that demonstrates a thorough understanding of the project and its intended output. Typically contains preliminary drawings and system descriptions etc. Together with the URS and the Detailed Design, it provides overall evidence that the design addresses the requirements of the equipment, system or facility.

Biocompatibility

A measure of how a biomaterial interacts in the body with the surrounding cells, tissues and other factors.

Bioburden

The level and type of micro-organisms that can be present in raw materials, API starting materials, intermediates or APIs. Bioburden should not be considered contamination unless the levels have been exceeded or defined objectionable organisms have been detected.

Biological Indicators
Test system containing viable microorganisms providing a defined resistance to a specified sterilization process, e.g. Vaporised hydrogen peroxide.

Biomaterial
Any matter, surface, or construct that interacts with biological systems. Biomaterials can be derived from nature or synthetic (manufactured). The active substance of a biosimilar medicine is comparable to a biological reference medicine. Biosimilar and biological reference medicines are used at the same dose to treat the same disease. The name, appearance and packaging of a biosimilar medicine differs to that of a biological reference medicine.

Bracketing

A Bracketing (aka family or matrix) approach can be used where similar products are produced using the same equipment and processes. A particular product size or product configuration may be selected to represent the worst-case product. Therefore, by qualifying the worst case, all of the other products within the family are considered validated.

Body orifice

Any natural opening in the body, as well as the external surface of the eyeball, or any permanent artificial opening, such as a stoma or permanent tracheotomy.

Borderline Classifications

In certain circumstances, it may not be clear if a product falls under the medical device legislation or whether to classify a device as a medicine, cosmetic, biocide and so on. The decision will largely depend on the particular intended use of the product, as assigned by the manufacturer, and on the demonstrated mode of action. The manufacturer's claims must be substantiated by relevant data.

Bulk Product

Any pharmaceutical form (liquid, powder, suspension) that is to be filled into either another container or its final container at the next process step; or is already filled into its final container to be labelled and packaged at the next process step.

BOM

Bill of Materials.

BSI

British Standards Institute.

C

CAD, Computer Aided Drawing

A system used to create physical designs, usually three-dimensional. Some examples of CAD software are SolidWorks, Pro/ENGINEER and AutoCAD.

Calibration

The a requirement that demonstrates a particular instrument or device produces results within specified limits by comparison with those produced by a reference or traceable standard over an appropriate range of measurements.

Campaign (Process)

A production strategy where consecutive batches of an API, a finished product, or intermediates are processed before the production line/system is cleaned.

Capability (Process Capability)

Process Capability is a measure of how capable the process is of producing product meeting specified requirements. It is a measure of the actual variation in that product characteristic compared to the product specifications. Indices are used to represent the Process Capability such as Pp, Cp and Ppk, Cpk, depending on how the data is collected e.g. multiple batches over time.

CAPA

A Corrective and preventive action. A systematic approach that includes actions needed to correct, prevent recurrence and eliminate the cause of potential nonconforming product and other quality problems (preventive action) (21CFR 820.100)

Change Control

A formal system by which qualified representatives of appropriate disciplines review proposed or actual changes that may impact the validated status.

Change Notification (Agreement)

A signed declaration that states that the Supplier agrees to notify the customer of changes in its product or process in order to allow the customer determine whether the changes can affect the quality of finished goods or Quality System.

Change Management

An overarching approach to change control that is used during the preliminary planning and design stage of a project.

Cleaning

The process of removing potential contaminants from process equipment and maintaining the condition of equipment such that the equipment can be safely used for subsequent product manufacture.

Cleaning Validation

Documented evidence that provides a high degree of assurance that a specific cleaning process will consistently produce a result meeting predetermined requirements for cleanliness.

Cleaning Verification

Confirmation by examination and provision of objective evidence that specific requirements have been fulfilled.

Cocurrent (flow)

This is when the fluids are applied in the same direction. Cocurrent flow is less effective as less heat can be transferred, therefore it is less commonly used.

Code of Federal Regulations (CFR)

Regulations issued by U.S. government agencies. The individual titles making up the regulations are numbered the same way as the federal laws on the same topic.

Competent Authority

A competent authority is the legally designated authority mandated to monitor compliance with directives and legal requirements within the industry. The competent authority has the power to grant and revoke licenses.

Compendial Organisations

Organizations certifying material standards that meet compendial requirements and acceptance criteria. (e.g. USP).

Commissioning

An engineering activity that includes all aspects of bringing a system, piece of equipment or process is installed and ready for use. Commissioning involves both requirements of Installation Qualification (IQ) and Operational Qualification (OQ).

Computer System

A group of hardware components and associated software, designed and assembled to perform a specific function or group of functions.
[EU GMP Guide, Part II, ICH Q7]

Computerised System

A system including the input of data, electronic processing and the output of information to be used either for reporting or automatic control. [EU GMP Guide, Glossary]

Computer System Validation

A process that confirms by examination and provision of objective evidence that the computer system conforms to user needs and intended uses. System validation is a process for achieving and maintaining compliance with GxP regulations and fitness for intended use by adoption of life cycle activities, deliverables, and controls.

Concurrent Validation

Concurrent Validation occurs when activities are executed at the same time as one another or concurrent to a product launch.

Confidence Level

Confidence Level is expressed as a percentage and represents the probability that the conclusion of the test is correct. A 95% confidence level means you can be 95% certain that the conclusion is correct.

Conflict Of Interest

A conflict of interest is a situation in which a public official's decisions are influenced by the official's personal interests.
☐
Continual Improvement, CI

Ongoing activities to evaluate and positively change products, processes, and the quality system to increase effectiveness

Consent Decree

A consent decree is a binding order issued by a judge that stipulates the voluntary agreement by the participants in a case of litigation. Decrees are sometimes issued after one party voluntarily agrees to cease a particular action without admitting to any illegality of the action to date.

Colony Forming Unit

One or more microorganisms that produce a visible, discrete growth on an agar-based microbiological medium.

Controlled Substances

Products that are categorized due to their potential for abuse, medical use and requirement for medical supervision.

Controlled classified areas

An environment supplied with HEPA filtered air where materials, equipment, and personnel are regulated to control viable and non-viable particulates to an acceptably low level. Such areas are classified according to the maximum level of airborne particulate allowed.

CNC (Controlled Not Classified)

While these are not ISO recognized room classes, they are generally used to describe non-GMP areas with a level of control in effect.

Clear (FDA)

"clear" the device after reviewing a premarket notification, otherwise known as a 510(k) (named for a section in the Food, Drug, and Cosmetic Act), that has been filed with FDA, or

Cleanroom

An area (or room or zone) with defined environmental control of particulate and microbial contamination, constructed and used in such a way as to reduce the introduction, generation and retention of contaminants within the area.

Containment

A process or device to contain product, dust or contaminants in one zone, preventing it from escaping to another zone.

Contamination

The undesired introduction of impurities of a chemical or microbial nature, or of foreign matter, into or onto a starting material or intermediate, during production, sampling, packaging or repackaging, storage or transport.

Continued Process Verification

Once the initial validation is completed it is important that the system or process remains within the validated state. This is done by monitoring the performance and output of the system or equipment. Furthermore, any changes to this system or equipment must be assessed and documented in order to assure the product is safe and meets acceptance criteria.

Critical Aspects

Critical aspects of manufacturing systems include the functions, features, abilities, and performance or characteristics required for the manufacturing process and systems to ensure consistent product quality and patient safety. They should be identified and documented based on scientific product and process understanding.
☐

Critical Quality Attribute, CQA (Critical-to-Quality)

A property or characteristic with specific nominal value and appropriate limit and range providing a particular quality attribute. A CQA typically is classed as a high risk requirement, where the safety or efficacy of the product depends on the CQA been within the specified limits.

CCC (Mark)

The "China Compulsory Certificate"mark, commonly known as CCC Mark, is a safety mark for many products sold on the Chinese market. As of 2013, medical devices do not require this certification.

CDC

Center for Disease Control & Prevention (USA)

CDRH

Center for Devices and Radiological Health (USA)

CE Marking

The CE Marking is a mandatory conformance mark on many products (including medical devices) placed on the single market in the European Economic Area. The CE marking certifies that a product has met EU consumer safety, health or environmental cquirements. By affixing the CE marking to a product, the manufacturer declares that it meets EU safety, health and environmental requirements.

CEN
Communité Européenne des Normes (European Committee for Standardization).

Clinical Trial

Clinical Trials are conducted to allow safety and efficacy data to be collected for health interventions (e.g., drugs, diagnostics, devices, therapy protocols). These trials can take place only after satisfactory information has been gathered on the quality of the non-clinical safety, and Health Authority/Ethics Committee approval is granted in the country where the trial is taking place.

Clinical Trial Sponsor

The Clinical Trial Sponsor is responsible for the safety of subjects in a clinical trial and informs local site investigators of the true historical safety record of the drug, device or other medical treatment to be tested, and of any potential interactions of the study treatment(s) with already approved medical treatments.

Cleaning

Removal of contamination or soils from an item or surface to the extent necessary for its further processing and its intended subsequent use.

CMDCAS

Canadian Medical Devices Conformity Assessment System.

CMDR

Canadian Medical Device Regulation.

Conformity

Fulfilment of a requirement or meeting a requirement.

Conformity Assessment Body (CAB)

A body, other than a Regulatory (competent) Authority, engaged in determining whether the relevant requirements in technical regulations or standards are fulfilled.

CRO

A "Contract Research Organization", also commonly known as a "Clinical Research Organization", is a service organization that provides support to the pharmaceutical and biotechnology industries. CROs offer clients a wide range of "outsourced" pharmaceutical research services to aid in the drug and medical device research and development process.

Data Integrity

Is the degree to which data is reliable and without error. Data must be accurate, attributable, contemporaneous, original, legible and available. A breach of data integrity occurs when any person manipulates or distorts data and submits the results of that data as valid.

Dead Leg

A dead leg in the world of piping terminology refers to an area of piping where there is insufficient flow or a tendency for water build-up or stagnation.
The formal definition of a dead-leg states that
Pipelines for the transmission of purified water for manufacturing or final rinse should not have an unused portion greater in length than 6 diameters (6D rule) of the unused portion of pipe measured from the axis of the pipe in use.

Debugging

The process of locating, analysing, and correcting suspected faults or machine issues. 333

Design controls

Design controls are a collection of practices and procedures that are incorporated into the design and development process for a product such as a medical device. It provides a structure and clear path from user needs assessment to product delivery through a step-by-step process. Design controls ensure proper assessment of the design is completed during the design and development phase. Design controls are a requirement of quality systems such as 21 CFR Part 820 (medical devices), and for certain classes of devices and per ISO 13485 - Quality Management Systems.

Decommissioning

When a system is taken out of production service and stored in an adequate environment for potential future use.

Depyrogenation

A thermal process used to destroy or remove pyrogens (endotoxins). Typically primary packaging components such as glass vials are subject to Depyrogenation.

Detection Limit

The lowest amount of analyte in a sample that can be detected but not necessarily quantitated as an exact value for an individual analytical procedure. (Ref: ICH Q2)

Design History File

The DHF is a repository for all of the documentation generated as a result of the design control process. The DHF serves as a complete record of the design.

Design Validation

Establishing by objective evidence that device or product specifications conform to user needs and intended use(s) defined in design documentation.

Debarment

The FDA has the authority to "disqualify," or remove, researchers from conducting clinical testing of new drugs and devices when the agency determines that the researcher has repeatedly or deliberately not followed the rules intended to protect study subjects and ensure data integrity. Further, the FDA can disqualify a clinical investigator who has repeatedly or deliberately submitted false information to the agency or study sponsor in a required report.
Under its statutory debarment authority, the agency may also ban, or "debar" from the drug industry individuals and companies convicted of certain felonies or misdemeanours related to drug products. Once individuals have been subjected to "debarment," they may no longer work for anyone with an approved or pending drug product application at FDA. Debarred companies may no longer submit abbreviated drug applications.

Design qualification (DQ)

The documented verification that the proposed design of the is suitable for the intended purpose. DQs are typical deliverables for facilities, systems and equipment and or processes.

Design Space

The multidimensional combination and interaction of input variables, e.g. material attributes, and process parameters that have been demonstrated to provide assurance of quality. Working within the design space is not considered as a change.

Directives

Directives are legal requirements. These must be met by manufacturers. Standard such as ISO 13485 help companies meet the requirements of directives, such as "Guidelines Relating to the Application of the Council Directive 93/42/EEC on Medical Devices."

Direct impact (system)

A system that is expected to have a direct impact on product quality. These systems are designed and commissioned in line with Good Engineering Practice (GEP) and, in addition, are subject to Qualification and Validation. Such systems include HVACs and Clean utilities such as WFI (Water-for-Injection)

Diffusion blending

A process in which particles are reoriented in relation to one another when they are placed in random motion and interparticular friction is reduced as the result of bed expansion (usually within a rotating container). Also referred to as tumble blending.

Deviations

A deviation can be simply described as an unintended event which causes a test or verification to fail to meet expected acceptance criteria.

Degree of invasiveness

A device, which in whole or in part, penetrates inside the body either through a body orifice or through the skin surface, is invasive. Invasiveness is generally categorised as invasive of a body orifice (including the surface of the eye), surgically invasive devices and implantable devices.

Device Master Record (DMR)

a compilation of records containing the procedures and specification for a device. The contents of a DMR can contain local procedures such as SOPs and work instructions along with global or divisional specifications used to detail manufacturing processes, intermediate product or final product.

Drug Product

The dosage form in the final immediate packaging intended for marketing. The finished dosage form that contains a Drug Substance, generally, but not necessarily in association with other active or inactive ingredients. (FDA)

Duration of Contact

In determining the classification of a device the duration that the device is in continuous contact with the patient is defined as transient, short term or long term. The longer the device is in contact with the patient or user, the greater the risk and therefore this has to be taken into account when determining classification. Continuous use is defined in MEDDEV 2.4/1 as the uninterrupted actual use for the intended purpose. Where use of a device is discontinued in order that the device is immediately replaced with an identical device (e.g. replacement of a urethral catheter) this shall be considered as continuous use of the device.

Electronic Signatures

Electronic signatures are computer-generated character strings that count as the legal equivalent of a handwritten signature. The regulations for the use of electronic signatures are set out in 21 CFR Part 11 of the FDA. Each electronic signature must be assigned uniquely to one person and must not be used by any other person. It must be possible to confirm to the authorities that an electronic signature represents the legal equivalent of a handwritten signature. Electronic signatures can be biometrically based or the system can be set up without biometric features.

Encapsulation

The division of material into a hard gelatin capsule. Encapsulators should all have the following operating principles in common: rectification (orientation of the hard gelatin capsules), separation of capsule caps from bodies, dosing of fill material/formulation, rejoining of caps and bodies, and ejection of filled capsules.

Endotoxin

A pyrogenic product (e.g., lipopolysaccharide) present in the bacterial cell wall. Endotoxin can lead to reactions in patients receiving injections ranging from severe fever to death.

Equipment Qualification

Qualification means the process to demonstrate the ability to fulfil specified requirements. EQ consists of proving and documenting that equipment or ancillary systems are properly installed (Installation Qualification, IQ), work correctly (Operations Qualification OQ), and the different sub-systems work together as a system (Performance Qualification PQ) and actually lead to the expected results.
Qualification is part of validation, but the individual qualification steps alone do not constitute a validated process.

Excipient

Substances other than the API which have been appropriately evaluated for safety and are intentionally included in a drug delivery system to provide a specific role in manufacturing, shelf-life or physical property.

Equipment Range

The full range that equipment is capable of performing, as per the manufacturer specification and tolerances. (a process may not utilize the full equipment range, operating over a narrower range).

F

Factory Acceptance Testing (FAT)

An FAT or Factory Acceptance Test is an engineering activity that inspects and verifies that the equipment or system meets the requirements of the URS.

Failure Mode And Effects Analysis (FMEA):

A risk assessment tool that provides for an evaluation of potential failure modes and their likely effect on outcomes and/or product or process performance in order to prioritize risks and monitor the effectiveness of risk control activities. It is often used to identify areas within a given process, product, or system that render it vulnerable.

FDA 483s

An FDA 483 letter typically includes a summary of findings and observations in relation to an audit or inspection where the FDA representatives have reason to believe GMP or other regulations have been violated or are not being met. In response to an FDA 483 letter, the company should address each item and provide a timeline for correction or request clarification of what changes are required.

Functional Design Specification (FDS)

A functional design specification is a document that specifies how particular requirements are met – this can be a combination of how the equipment/process operates mechanically/automatically etc. An FDS is typically written to response to a URS

Fluid

A fluid is a substance that undergoes continuous deformation when subjected to a shearing force.

☐

☐

G

GAMP

Good Automated Manufacturing Practice (GAMP) is a set of guidelines for manufacturers and users of automated systems in regulated industries. Specifically, the Medical device, pharmaceutical and biopharmaceutical industries.
The application of GAMP and Validation of Automated Systems in manufacturing helps ensure that regulated medical devices and medicinal products have the required quality and are manufactured according to Good practices, meet regulatory and legal requirements and ensure patient safety.

Good Documentation Practices, GDP

The handling of written or pictorial information describing, defining, specifying and/or reporting of certifying activities, requirements, procedures or results in such a way as to ensure data integrity

Granulation

A process of creating granules. The powder morphology is modified through the use of either a liquid that causes particles to bind through capillary forces or dry compaction forces.

Grade A Areas

Aseptic processing areas, critical in nature where sterile products are exposed to the environment receiving no further sterilization. High-risk operations (for example aseptic stopperage, filling, loading of the lyophilizer) occur in Grade A areas. They are considered ISO 5 under both dynamic and static conditions.)

Grade B Areas

Aseptic processing areas where the sterile product is protected from the environment. Grade B processing areas are the background environments for Grade A areas and are considered ISO 7 environments in the dynamic state and ISO 5 environments under static conditions.

Grade C Areas

Non-critical areas where bulk product or materials are exposed to the environment, yet final sterilization has not yet been performed. Grade C areas are support areas for non-sterile production activities; purification, formulation, and preparation of components, equipment, etc. for sterilization. They are considered ISO 8 (Class 100,000) environments in the dynamic state and ISO 7 (Class 10,000) environments under static conditions.

Grade D Areas

Non-critical production areas, support areas, airlocks, or corridors. They are support areas for non-sterile production activities in closed systems; cell culture, or buffer and media preparation areas. Grade D Airlocks are used for the movement of product, materials, and personnel into classified areas.

GHTF

Global Harmonization Task Force

GxP

GxP is a general term for good practice with regard to quality guidelines and regulations. These guidelines are used in many fields, including the pharmaceutical, medical device and food industries. "x" is used as an umbrella letter representing different subjects or disciplines in industry. Some prime examples include GLP (Good Laboratory Practice), GDP (Good Documentation Practice), GEP (Good Engineering Practice) and GMP (Good Manufacturing Practices). Furthermore, the use of a lower case "c" as a prefix indicates "current" or "up-to-date"

H

Harm

Damage to health, including the damage that can occur from loss of product quality or availability.

High level risk assessment (HLRA)

A High level risk assessment that can be used at the beginning of a project to estimate the risk. Such as the risks involved with bringing in new computerised/automated equipment.

HVAC

Heating, ventilation and air-conditioning (HVAC) systems are used to control the environmental conditions within an area or manufacturing facility. HVAC systems also provide comfortable conditions for operators based in the manufacturing environment. Temperature, relative humidity (RH) and ventilation should not adversely affect the quality of products during their manufacture and storage, or the proper functioning of equipment

Hydrogel

A biomaterial made up of a network of polymer chains that are highly absorbent and as flexible as natural tissue.

I

ICH

International Conference on Harmonization of Technical Requirements for Registration of Pharmaceuticals for Human Use.

Intended Purpose

Intended purpose means the use for which the device is intended according to the data supplied by the manufacturer on the labelling, in the instructions and/or in promotional materials. (Chapter I section 1 of Annex IX of Directive 93/42/EEC)

Impurity

Any component of the new active pharmaceutical ingredient which is not the chemical entity defined as the new active pharmaceutical ingredient OR any component present in the active pharmaceutical ingredient or final product which is not the desired product, a product-related substance, or excipient including buffer components.

Invasive device

A device, which, in whole or in part, penetrates inside the body, either through a body orifice or through the surface of the body.

IQ/OQ

Equipment IQ/OQ is defined as establishing documented evidence that all key aspects of the process equipment installation adhere to the manufacturer's approved specifications and any recommendations of the supplier of the equipment are suitably considered.
The process/equipment must also operate as intended and all user requirements are adequately fulfilled.

IFU

Instructions for Use.

(Plant) Injunction

An injunction is a judicial process initiated to stop or prevent violation of the law, such as to halt the flow of violative products in interstate commerce and to correct the conditions that caused the violation to occur. (FDA 21 U.S.C. 332; Rule 65, Rules of Civil Procedure).
If a firm has a history of violations and has promised correction in the past but has not made the corrections, the injunction is more likely to succeed. However, the freshness of the evidence is critical.

For an injunction action to be credible in the eyes of the Department of Justice (DOJ), the U.S. Attorney and the court, the evidence must be current. Timeliness is an important factor when considering an injunction action, with or without a Motion for Preliminary Injunction or a temporary restraining order (TRO). However, case quality and credibility must not be sacrificed to meet guideline time frames. The purpose of the guideline time frames is to limit, as much as can reasonably be expected, the need to update evidence. Updating entails extra work at all levels of the case development and review process and more importantly, delays obtaining an injunction which is intended to stop violations that adversely affect the safety or quality of products in commerce.

ISO

International Organization for Standardization. Agency responsible for developing international standards. E.g. ISO 13485 Medical Devices.

Isolator
A sealed enclosure, which provides full physical separation between the critical processing zone and the surrounding other processing zones. The internal surfaces of the isolator and of its contents are decontaminated, in accordance with defined objectives, by highly effective cycles. (e.g. Vaporised Hydrogen peroxide) Enclosure capable of preventing ingress of contaminants by means of physical interior/exterior separation, and capable of being subject to reproducible interior bio-decontamination.

Isoelectric Precipitation

Isoelectric Precipitation works by reducing the electrostatic forces to near zero, allowing the proteins to precipitate out.

ISO 13485

ISO 13485, ISO standard, published in 2003, that represents the requirements for a comprehensive management system for the design and manufacture of medical devices.

ISO 14971

An ISO standard, published in 2007, that provides a framework and requirements for a risk management system for medical devices. This standard establishes the requirements for risk management to determine the safety of a medical device by the manufacturer during the product life cycle.

ISO 9001

ISO 9001 is an ISO standard that represents the requirements for quality management systems. It is used across industries and is not specific to medical devices like ISO 13485.

Item Master

A of all components that a manufacturer buys, builds or assembles into its products. The item master includes information like the size, shape, material, manufacturer, manufacturer part number and vendor for each component.

IVD

In Vitro Diagnostic tests are medical devices intended to perform diagnoses from assays in a test tube, or more generally in a controlled environment outside a living organism.

IVDD

The In Vitro Diagnostic Device Directive delineates requirements that in vitro diagnostic devices must meet before they can be sold in the EU market.

Intermediate

A material produced during steps of the processing of an API that undergoes further molecular change(s) or purification before it becomes an API.

J

JIT (Just in time)

A strategy used to monitor inventory levels with the goal of reducing inventory and associated carrying costs.

K

Kanban

A scheduling system that advises manufacturers what to produce, when to produce and how much to produce. Pioneered by Toyota, the approach is based on demand. Inventory is replenished only when visual cues like an empty bin, trolley or cart show that it's needed.

L

Laminar flow

Laminar flow is when fluid particles move in parallel layers, at a constant velocity.

Lifecycle (Validation)

The Validation lifecycle refers to the requirement to control and document all validation activities from conception and URS stage to the retirement of equipment or a process. The lifecycle approach ensures compliance throughout the life of the process/equipment while maintaining a validated state throughout the application of change control.

Linearity

The ability of an analytical procedure (within a given range) to obtain test results that are directly proportional to the concentration (amount) of analyte in the sample.

Line Clearance

The act of performing and documenting the removal of materials from a production or packaging line and cleaning prior to the introduction of a new batch or lot.

Lyophilization (or Freeze Drying)

Lyophilization is the removal of ice or other frozen solvents from a material through the process of sublimation and the removal of bound water molecules through the process of desorption.

M

Maximum Allowable Carry Over (MACO)

The amount of allowed product residue (carry-over) from lot-to-lot, batch-to-batch. This limit is based on the most conservative or lowest level of three MACO calculation methods (1) Limited based on Toxicity, (2) Limit based on Smallest Therapeutic Dose, and (3) Worst Case Dose.

Measurement Capability Index (MCI)

The Measurement Capability Index (MCI) represents the capability of the measurement system. It is used to evaluate the capability of the gauge to classify product against predetermined specifications.

Measurement System Analysis (MSA)

A study to determine the degree of error involved in measuring the given parameter. The measurement system involves the combination of operations, procedures, gauges, instruments, environmental conditions, people and software.

Medical Device

A medical device is "an instrument, apparatus, implement, machine, contrivance, implant, in vitro reagent, or other similar or related article, including a component part, or accessory which is:

• recognized in the official National Formulary, or the United States Pharmacopoeia, or any supplement to them,
• intended for use in the diagnosis of disease or other conditions, or in the cure, mitigation, treatment, or prevention of disease, in man or other animals, or
• intended to affect the structure or any function of the body of man or other animals, and which does not achieve any of its primary intended purposes through chemical action within or on the body of man or other animals and which is not dependent upon being metabolized for the achievement of any of its primary intended purposes."

Medicinal Drug Products (Finished Products)

Finished dosage forms (e.g. tablet, capsule, or solution) that contain the active pharmaceutical ingredient usually combined with inactive ingredients. Medicinal products are intended to furnish pharmacological activity or other direct effect in the diagnosis, cure, mitigation, treatment, or prevention of disease or to affect the structure and function of the body.

MDD

The Medical Device Directive is intended to harmonize the laws relating to medical devices within the European Union. Medical Device Directive 93/42/EEC was most recently reviewed and amended by 2007/47/EC.

MHRA

The Medicines and Healthcare products regulatory Agency (MHRA) is the UK government agency which is responsible for ensuring that medicines and medical devices work and are acceptably safe.

MSDS

Material Safety Data Sheet.

N

NCR

Non-Conformance Report.

NIH

National Institutes of Health (U.S.)

Noel

No Observed Effect Level. In relation to Cleaning Validation.

Non-conformity

A deficiency in a characteristic, product specification, CQA, process parameter, record, or procedure that renders the quality of a product unacceptable, indeterminate, or not according to specified requirements.

Non Parametric Data

Where the type of data is non variable Also referred to as attribute data eg (Visual inspection resulting in a PASS/FAIL result.

Notified Bodies

A notified body is a certification organisation which the national authority (the competent authority) of a member state designates to carry out one or more of the conformity assessment procedures or audits described in the annexes of the medical devices directives or GMP legislation.

NPI (New product introduction)

The market launch or commercialization of a new product. NPI takes place at the end of a successful product development project.

⬜

O

Open System

An environment in which system access is not controlled by persons who are responsible for the content of electronic records on the system (21 CFR, Part 11)

Outlier

A test result that is statistically different compared to a set of other test results obtained from the same sample or samples from the same lot of material.

Out-Of-Specification

A recorded result that falls outside the established specification(s) or acceptance criteria.

Out-Of-Trend

Analytical result, which is within specification or acceptance criteria, but different from those usually obtained or expected. Out-of-trend results should be investigated by the same general principles as out-of-specification results.

Quantitation limit

The lowest amount of analyte in a sample which can be quantitatively determined with suitable precision and accuracy for an analytical procedure. The quantitation limit is a parameter of quantitative assays for low levels of compounds in sample matrices and is used particularly for the determination of impurities and degradation products.

Overall Equipment Effectiveness(OEE)

A calculation for measuring the efficiency and effectiveness of a process, by Equipment breaking it down into three constituent components (the OEE Factors) Availability x Performance x Quality.

Overkill

Sterilization process that is demonstrated as delivering at least a 12 Spore Log Reduction (SLR) to a biological indicator having a resistance equal to or greater than the bioburden level.

136

☐

P

Pan Coating

The uniform deposition of coating material onto the surface of a solid dosage form while being translated via a rotating vessel.

Particle count test

Test covers verification of cleanliness. Dust particle counts measured. The number of readings and positions of tests should be defined in accordance with ISO 14644-1 Annex B5.

Performance indicators

Measurable values used to quantify quality objectives to reflect the performance of an organization, process or system, also known as performance metrics in some regions. (ICH Q10)

Performance Qualification (PQ)

Establishing by documented evidence that the process, under anticipated (controlled) conditions, consistently produces a product which meets predetermined requirements.

Precision

The degree of agreement (scatter) between a series of measurements when a method is applied repeatedly to multiple samplings of a homogeneous sample or artificially prepared sample under the prescribed conditions. There are three types of precision; repeatability, intermediate precision and reproducibility.

Pressure cascade

A process whereby air flows from one area, which is maintained at a higher pressure, to another area at a lower pressure.

Piping & Instrument Diagrams (P&IDs)

Engineering technical drawings that provide details of the connections and integration of equipment, services, material flows, plant controls and alarms. The P&ID also provide the reference for each tag or label used for identification.

PMA

Premarket approval by FDA is the required process of scientific review to guarantee safety and effectiveness for Class III devices.

PMDA

The Pharmaceutical and Medical Devices Agency in Japan reviews applications for marketing approval of pharmaceuticals and medical devices. It also monitors their post-marketing safety and provides relief compensation for people who have suffered from adverse drug reactions from pharmaceuticals or infections from biological products.

PMS

Post Marketing Surveillance is the practice of monitoring a pharmaceutical drug or device after it has been released on the market.

Process design

Defining the commercial manufacturing process based on knowledge gained through development and scale-up activities.

☐
Process qualification

Confirming that the manufacturing process as designed is capable of reproducible commercial manufacturing.

Process window

The selected operating range of machine setting/parameter that will produce product to meet all quality and product specifications.

Product Recovery

Product recovery is a critical and important step in the process. It is also referred to as "Downstream processing". It is often the most expensive step in the process. For recombinant-DNA derived products, purification can often account for 90% of the total production costs.

Prospective Validation

Prospective Validation is when validation is done in advance of commercial manufacturing.

Procedures

Also known as Standard Operating Procedures, or SOPs, give directions for performing certain operations.

Protocols

Give instructions for performing and recording certain discreet operations. (Examples include engineering protocols, validation protocols etc.)

Pure

A term typically used within pharmaceutical manufacturing, a product or substance is pure if it is free of contaminants, foreign matter, chemicals and harmful microbes.

☐

Q

QMS

Quality Management System can be expressed as the organizational structure, procedures, processes and resources needed to implement quality management.

Quality

The degree to which a set of inherent properties of a product, system, or process fulfils requirements. (ICH Q9)

Quality by design

This is a systematic approach that begins with predefined objectives and emphasizes product and process understanding and process control, based on sound Science and engineering principles.

Quality Management System

A Quality Management System, often abbreviated to (QMS) is any system based on a collection of business processes that are primarily focused on providing safe and quality products that consistently meet customer requirements.

☐
Quarantine

The status of materials isolated physically or by other effective means pending a decision on their subsequent approval or rejection.

(Quality) Policy

A document in which a company or organization outlines their commitment and approach to quality. It usually sets out how they plan to achieve a high and consistent standard of quality. It should in some way speak to the customer or end user.
☐
Qualification Plan

A Qualification Plan (QP) describes all the qualification measures and at which stage of the qualification the verification will be completed. It typically contains detailed descriptions of the necessary test measures and a description of the interdependencies of the individual tests. In some instances, there may not be a need or a requirement for a qualification plan. A validation plan can also serve to detail the qualification strategy.

QP

Companies that intend to manufacture or import medicinal products or intermediate products, for use in clinical trials or for market within the EU, must appoint the service of a Qualified Person, in order to comply with EU Good Manufacturing Practice Standards.

QPM

Quality Policy Manual.

QSP

Quality System Procedure.

QSR

Quality System Regulations.

R

Range

Range is defined as the interval between the upper and lower measurements required. The minimum specified range should be within the equipment range and validated to operate at all points within the range.

Recall

As defined at 21 CFR 7.3(g), "recall means a firm's removal or correction of a marketed product that the Food and Drug Administration considers to be in violation of the laws it administers and against which the agency would initiate legal action, 2 21 CFR 806.2(h). e.g., seizure. Recall does not include a market withdrawal or a stock recovery." Recall does not include routine servicing. Recall also does not include an enhancement, as defined by this guidance.

Relative humidity

The ratio of the actual water vapour pressure of the air to the saturated water vapour pressure of the air at the same temperature expressed as a percentage. More simply put, it is the ratio of the mass of moisture in the air, relative to the mass at 100% moisture saturation, at a given temperature.

Reusable medical device

A device intended for repeated use either on the same or different patients, with appropriate decontamination and other reprocessing prior to re-use.

Reusable Surgical Instrument

Instrument intended for surgical use by cutting, drilling, sawing, scratching, scraping, clamping, retracting, clipping or similar surgical procedures, without connection to any active medical device and which are intended by the manufacturer to be reused after appropriate procedures for cleaning and/or sterilisation have been carried out.
☐
Re-Qualification

Requalification is designed to verify and ensure that the equipment/instrument/system is maintained in a qualified state after modification or after a stipulated time period (downtime).

Residual Risk

The risk level remaining after applying the identified controls on a high risk of harms and hazards manifestation.

Resolution

The smallest change in quantity that can be detected or provided by an instrument.

Residual Solvent
Organic volatile chemicals used or produced during the manufacture of APIs or excipients, or in the preparation of medicinal products.

Retain Samples
Samples that are kept for potential investigations and retests. It should be noted that retain samples are not a regulatory requirement, per Annex 10 or 21 CFR part 11.

Retrospective Validation

Retrospective validation is used for facilities or processes that have not completed formal validation. Historical data or a retrospective review can provide the evidence that the process or facility is operated as intended.

Rinse Sampling

Using a solvent to contact all surfaces of the sampled item to quantitatively remove target residue. The solvent can be water, water with pH adjusted, or organic solvent.

Right First Time

Right First Time strives to create a culture of excellence. People are challenged with performing their tasks always in the correct manner to achieve the correct results always - right the first time.

Risk

The combination of the probability of occurrence of harm and the severity of that harm.

Risk Management

Risk management involves the systematic application of management policies, practices and procedures that identify, analyse, control and monitor risk.
It is important to recognise that risk management should begin at the outset of the design and development phase of a project. The first step is to identify the user needs and intended use and application of the device.

RoHS

"Restriction of Hazardous Substances in electrical and electronic equipment 2002/95/EC". An initiative that was adopted by the European Union (EU) in February 2003 and put into effect July 1, 2006
Ruggedness

An indication of how resistant a test method or process is to typical variations in operation, such as those to be expected when using different analysts, different instruments and different reagent batches.
☐
☐

S

Scaffold

A structure of artificial or natural materials on which tissue is grown to mimic a biological process outside the body.

SKU

(Stock keeping unit) A unique sales stock identifier.

Specifications

A approved document detailing the requirements with which the products or materials used or obtained during manufacture have to conform. They serve as a basis for quality evaluation.

Specificity

The ability to assess unequivocally the analyte in the presence of components, which may be expected to be present.

Stability

Stability studies are used to demonstrate and justify assigned expiration or retest dates.

5S

5S is a Japanese methodology of organising and storing items in a work or lab environment. It has been adopted by many Western companies as a tool to help maintain standards and reduce errors and mix-ups. The "5s" represents each stage of the method.
Sort
Sorting out any items that are not in use and removing to a more appropriate area or to storage or the bin.
Set-in-Order
The idea of "Set-in-Order" is to be always organised. "A place for everything and everything in its place. "If we "set-in-order" we can help to make live processing and testing more efficient and reduce the risk of errors, omissions and accidents.
Shine
Regular cleaning is an important practice and it is always helpful to "Clean as you go."
Standardise
Implement standard practices through SOPs and training. Standardisation can also be applied to work station layout.
Sustain

Make it a habit! After implementing a 5s methodology, it is only effective if continuous efforts are made to "sustain" the changes.

Sterility Assurance (SAL)

Probability of a single viable microorganism occurring on an item after sterilization. For a terminally sterilized medical device to be designated as "sterile", the minimum sterility assurance level shall be SAL = 10-6 or better. When applying this quantitative value to assurance of sterility, an SAL of 10-7 has a lower value but provides a greater assurance of sterility than an SAL of 10-6 .

☐

T

Tableting

The reconstitution of a powder blend in which compression force is applied to form a single unit dose. (tablet)

Tableting press

Tablet press subclasses primarily are distinguished from one another by the method that the powder blend is delivered to the die cavity. Tablet presses can deliver powders without mechanical assistance (gravity), with mechanical assistance (automation), by rotational forces (centrifugal), and in two different locations where a tablet core is formed and subsequently an outer layer of coating material is applied (compression coating).

Traceability Matrix

A Traceability Matrix is a document that links the user requirements and specifications to where the verification and testing has been documented within the validation activities. A traceability matrix illustrates that all user requirements are traceable to the evidence based test.

Turbulent flow

Turbulent flow is when the movement of fluid particles are varying in velocity and direction.

U

Uniform

The product is manufactured consistently and will have the same quality between batches manufactured on different days.

UDI, Unique Device Identification

The UDI is a series of numeric or alphanumeric characters that is created through a globally accepted device identification and coding standard. It allows the unambiguous identification of a specific medical device on the market.

Uninterrupted Power Supply

An uninterruptible power supply (UPS) is a system for buffering the main power supply. If the power supply fails, the battery of the UPS supplies the required power. When the power supply returns, the UPS battery stops supplying power and is recharged.

Unit Operation

Unit operations are the individual steps in the process that modify materials and their properties at each step of the process. Each unit operation comes together to create a complete process.

User Requirement Specification, URS

The URS is a critical document that defines the requirements of a particular system, equipment or process. Requirements such as the functional and operational aspects of the system are typically documented here.

USP

United States Pharmacopoeia.

V

Validation

Validation is confirmation via documented evidence that the particular requirements for a specific intended use can be consistently fulfilled under anticipated conditions.

Validation Master Plan

A document providing information on a company's validation work programme. It typically details timescales for the validation work to be performed along with the key deliverables.

Verification

Verification confirmation by examination and provision of objective evidence (i.e. documentation) that the specified requirements have been fulfilled.

Vaporized Hydrogen Peroxide (VHP)

Vaporization of liquid hydrogen peroxide which results in a mixture of VHP and water vapor. The VHP mixture is used to decontaminate isolators.

W

Warning Letter

A warning letter is a correspondence that notifies regulated industry about violations that FDA has documented during its inspections or investigations.

WEEE Directive

Waste electrical and electronic equipment directive. European Community directive 2002/96/EC where manufacturers are responsible for disposing of electrical/electronic waste.

WFI (Water for injection)

WFI is sterile and pyrogen free water containing o less than 10 CFU/100ml (Colony Forming Units) with a sample size of between 100 and 300 ml and an endotoxin level < 0.25 EU/ml.

WHO

World Health Organization.

WI

Work Instructions.

Witnessed By

When signed or initialed is legal proof that the individual signing is physically present and observes the step, calculation, or operation being performed by someone else, and that all entries of data are true and accurate.

Worst Case

A set of conditions or parameters which, in combination with product specification or attributes at their limits, pose the greatest challenge to the process.

X

--

Y

--

Z

Zone Classification

Zone classification refers to GMP areas which include controlled (aka classified) and non-controlled manufacturing areas. Areas may be classified based on EU Grades A–D and/or ISO Class 5–8 (in the US - Class 100–Class 100,000 areas.

www.ingramcontent.com/pod-product-compliance
Lightning Source LLC
Chambersburg PA
CBHW081559220526
45468CB00010B/2700